GRAPHING CALCULATOR AND EXCEL® MANUAL

FLORENCE CHAMBERS
Southern Maine Community College

COLLEGE ALGEBRA IN CONTEXT WITH APPLICATIONS FOR THE MANAGERIAL, LIFE, AND SOCIAL SCIENCES

SECOND EDITION

Ronald J. Harshbarger
University of South Carolina – Beaufort

Lisa S. Yocco
Georgia Southern University

PEARSON
Addison Wesley

Boston San Francisco New York
London Toronto Sydney Tokyo Singapore Madrid
Mexico City Munich Paris Cape Town Hong Kong Montreal

Reproduced by Pearson Addison-Wesley from electronic files supplied by the author.

Publishing as Pearson Addison-Wesley, 75 Arlington Street, Boston, MA 02116.

ISBN 0-321-36976-9

3 4 5 6 OPM 09 08 07

PEARSON
Addison
Wesley

CHAPTER 1 FUNCTIONS, GRAPHS, AND MODELS **1**

GRAPHING WITH TECHNOLOGY - TI-83, TI-83 PLUS, AND TI-84 PLUS 1
Getting started *1*
Preliminaries to graphing on the TI-83, TI-83 Plus, and TI-84 PLUS 1
Input the function $y = -x^2 + 2x - 1$ 1
Window Settings 1
Setting the Standard Window [-10, 10, 1] by [-10, 10, 1] 2
Graph the function(s) 2
Trace the function 2
Display the equation on the graph 2
Input the function $y = x^3 - 3x^2 - 13.$ 3
Turn functions on and off 3
Set the viewing window for Y2 = $x^3 - 3x^2 - 13.$ 3
When more than the Graph is displayed 4
When you get an Invalid Dim Error message 4

GRAPHING WITH TECHNOLOGY - TI- 86 5
Getting started *5*
Preliminaries to graphing on the TI-86 5
Input the function $y = -x^2 + 2x - 1$ 5
Window Settings 5
Setting the Standard Window [-10, 10, 1] by [-10, 10, 1] 6
Graph the function(s) 6
Trace the function 6
Display the equation on the graph 7
Input the function $y = x^3 - 3x^2 - 13.$ 7
Turn functions on and off 7
Set the viewing window for y2 = $x^3 - 3x^2 - 13.$ 8
When more than the Graph is displayed 9
When you get an Invalid Dim Error message 9

1.2 EXAMPLE 2 ON THE TI-83, TI-83 PLUS AND TI-84 PLUS 10

Graph the Function $C = \dfrac{5350p}{100-p}$ *10*

Enter the Equation 10
Set the Window 10
Graph the equation 10
Find $x = 90$ directly on the graph 10
Find $x = 90$ using the TABLE ASK feature 11
Find $x = 90$ by making a TABLE of values 11

1.2 EXAMPLE 2 ON THE TI-86 12

Graph the Function $C = \dfrac{5350p}{100-p}$ *on the TI-86* *12*

Enter the Equation 12
Set the Window 12
Graph the equation 12
Find $x = 90$ directly on the graph 13
Find $x = 90$ using the TABLE ASK feature 13
Find $x = 90$ by making a TABLE of values 13

1.2 EXAMPLE 2 USING EXCEL 14

Graph the Function $C = \dfrac{5350p}{100-p}$ *in EXCEL* *14*

Enter the Data Set 14
Format column B to currency notation 15
Graph the Equation 16

1.2 EXAMPLE 5 ON THE TI-83, TI-83 PLUS, AND TI-84 PLUS 18
Create a Scatter Plot of Data Points on the TI-83, TI-83 Plus, and TI-84 Plus *18*
Set the Window 18
Enter the Data Points 18
Clearing Lists 19
Turn on a Plot 20
Use STAT PLOT to format the graph of data points 20

1.2 EXAMPLE 5 ON THE TI-86 21
Create a Scatter Plot of Data Points on the TI-86 ..... *21*
Set the Window 21
Enter the Data Points 21
Clearing Lists 22
Turn on a Plot 22
Use STAT PLOT to format the graph of data points 23

1.2 EXAMPLE 5 USING EXCEL 24
Create a Scatter Plot of Data Points in EXCEL *24*
Enter the Data Set 24
Graph the Scatter Plot 25

1.5 EXAMPLE 5 ON THE TI-83, TI-83 PLUS, AND TI-84 PLUS 27
Solve a Linear Equation Graphically Using the x-Intercept Method. *27*
Enter the Equation $\frac{2x-3}{4}=\frac{x}{3}+1$ 27
Set the Window 27
Graph the equation and use TRACE to estimate the zero 27
Find the exact x-intercept 28
Solve a Linear Equation Graphically Using the Intersection Method on the TI-83, ... *29*
TI-83 Plus, and TI-84 PLUS. *29*
Enter the Equation $\frac{2x-3}{4}=\frac{x}{3}+1$ 29
Set the Window 29
Graph the equations 29
Find the exact coordinates of the Intersection point 30

1.5 EXAMPLE 5 ON THE TI-86 31
Solve a Linear Equation Graphically Using the x-Intercept Method on the TI-86 .... *31*
Enter the Equation $\frac{2x-3}{4}=\frac{x}{3}+1$ 31
Set the Window 31
Graph the equation 31
Find the exact x-intercept 32
Find the exact x-intercept 32
Solve a Linear Equation Graphically Using the Intersection Method on the TI-86 .... *33*
Enter the Equation $\frac{2x-3}{4}=\frac{x}{3}+1$ 33
Set the Window 33
Graph the equations 33
Find the exact coordinates of the Intersection point 34

1.5 EXAMPLE 6 ON THE TI-83, TI-83 PLUS, AND TI-84 PLUS 35

1.5 EXAMPLE 6 ON THE TI-86 35

1.5 EXAMPLE 6 USING EXCEL 35
Find the x-intercept of a graph using GOAL SEEK in EXCEL. *35*
Find the exact coordinates of the Intersection point using GOAL SEEK in EXCEL. 37

1.6 EXAMPLE 1 ON THE TI-83, TI-83 PLUS, AND TI-84 PLUS 38
Create the scatter plot - See 1.2 Example 5 for detailed instructions. *38*
Find the Linear Regression Equation that is the "Best Fit" for the Data. *39*
Display r and r^2 .... *39*
Graph the Linear Regression Equation on the Graph Screen. *40*

1.6 EXAMPLE 1 ON THE TI-86 41
Set up the STAT editor to include L1, L2, ..., L6. .... *41*
Create the scatter plot - See 1.2 Example 5 for detailed instructions. *41*
Find the Linear Regression Equation that is the "Best Fit" for the Data. *42*
Graph the Linear Regression Equation on the Graph Screen. *43*

1.6 EXAMPLE 1 USING EXCEL.......45
Create the scatter plot - See 1.2 Example 5 for detailed instructions........45
Graph the three lines on the scatter plot.......47
Find the regression equation and put it on the scatter plot........49

1.7 EXAMPLE 2 ON THE TI-83, TI-83 PLUS, AND TI-84 PLUS.......50
Solve a system of linear equations by graphing........50
Find the exact coordinates of the Intersection point.......50

1.7 EXAMPLE 2 ON THE TI-86.......51
Solve a system of linear equations by graphing........51
Find the exact coordinates of the Intersection point.......51

1.7 EXAMPLE 2 USING EXCEL.......52
Solve a system of linear equations by graphing........52
Find the exact coordinates of the Intersection point using GOAL SEEK in EXCEL.......53

CHAPTER 2 QUADRATIC AND OTHER NONLINEAR FUNCTONS.......55

2.1 EXAMPLE 1 ON THE TI-83, TI-83 PLUS, AND TI-84 PLUS.......55
Vertices of Parabolas.......55
Display the vertex on the Graph of the function $P(x) = -2x^2 + 60x$.......55

2.1 EXAMPLE 1 ON THE TI-86.......56
Vertices of Parabolas.......56
Display the vertex on the Graph of the function $P(x) = -2x^2 + 60x$.......56

2.2 EXAMPLE 5 ON THE TI-83, TI-83 PLUS, AND TI-84 PLUS.......57
Marijuana Use.......57
Use the Intersection Method to solve $33 = -0.2286t^2 + 2.7783t + 49.8138$.......57
- See 1.5 Example 5 for more details of this method.......57
Use a Table to find the solution.......57

2.2 EXAMPLE 5 ON THE TI-86.......58
Marijuana Use.......58
Use the Intersection Method to solve $33 = -0.2286t^2 + 2.7783t + 49.8138$.......58
- See 1.5 Example 5 for more details of this method.......58
Use a Table to find the solution.......58

2.3 GRAPHING MORE NONLINEAR FUNCTIONS ON THE TI-83, TI-83 AND TI-84 PLUS 59
Power Functions such as $C = 0.11W^{1.54}$ *of Example 1* 59
Root Functions such as $y = x^{2/3}$ 59
Reciprocal Functions Graph $y = \dfrac{1}{x+5}$ 60
Change the MODE to DOT 60
Graph the function 60
Graph a Piece-Wise Function $f(x) = \begin{cases} 5x+2 & \text{if } 0 \le x < 3 \\ x^3 & \text{if } 3 \le x \le 5 \end{cases}$ *of Example 3* 60
Absolute Value Function $|x| = \begin{cases} x & \text{if } x \ge 0 \\ -x & \text{if } x < 0 \end{cases}$ 61

2.3 GRAPHING MORE NONLINEAR FUNCTIONS ON THE TI-86 62
Power Functions such as $C = 0.11W^{1.54}$ *of Example 1* 62
Root Functions such as $y = x^{2/3}$ 62
Using the caret key ^ 62
Using $\sqrt[x]{\ }$ in the Math Menu 62
Reciprocal Functions Graph $y = \dfrac{1}{x+5}$ 63
Format the Graph to DrawDot 63
Graph the function 63

2.5 EXAMPLE 1 ON THE TI-83, TI-83 PLUS, AND TI-84 PLUS 64
Create the scatter plot - See 1.2 Example 5 for detailed instructions. 64
Find the Quadratic Equation to Model the Data. 64
Graph the Quadratic Regression Equation on the Graph Screen 65

2.5 EXAMPLE 1 ON THE TI-86 66
Create the scatter plot - See 1.2 Example 5 for detailed instructions. 66
Find the Quadratic Equation to Model the Data 67
Graph the Quadratic Regression Equation on the Graph Screen 67

2.5 EXAMPLE 1 USING EXCEL 68
Create the scatter plot - See 1.2 Example 5 for detailed instructions. 68
Find the regression equation and put it on the scatter plot. 68
Evaluate a Formula 69

2.5 EXAMPLE 4 ON THE TI-83, TI-83 PLUS, AND TI-84 PLUS 70
Create the scatter plot 70
Find and graph the power function model for this data 70

2.5 EXAMPLE 4 ON THE TI-86 71
Create the scatter plot 71
Find and graph the power function model for this data 71

2.5 EXAMPLE 4 USING EXCEL 72

2.6 EXAMPLE 7 ON THE TI-83, TI-83 PLUS, AND TI-84 PLUS 73
Graph $(f+g)(x)$ in the window [-1, 10, 1] by [0, 200, 50] 73
Graph $(f/g)(x)$ in the window [-1, 10, 1] by [0, 0.017, .001]. 74
Compute $(f \circ g)(-3)$ and $(g \circ f)(9)$ 74

2.6 EXAMPLE 7 ON THE TI-86 75
Graph $(f+g)(x)$ in the window [-1, 10, 1] by [0, 200, 50] 75
Graph $(f/g)(x)$ in the window [-1, 10, 1] by [0, 0.017, .001]. 75
Compute $(f \circ g)(-3)$ and $(g \circ f)(9)$ 76
Put y1 and y2 into CUSTOM 76

CHAPTER 3 EXPONENTIAL AND LOGARITHMIC FUNCTIONS 77

3.4 EXAMPLE 2 ON THE TI-83, TI-83 PLUS AND TI-84 PLUS 77
Create the scatter plot. 77
Find the Exponential Equation to Model the Data. 77
Graph the Exponential Regression Equation on the Graph Screen 78

3.4 EXAMPLE 2 ON THE TI-86 79
Create the scatter plot 79
Find the Exponential Equation to Model the Data. 79
Graph the Exponential Regression Equation on the Graph Screen 80

3.4 EXAMPLE 2 USING EXCEL 81
Create the scatter plot 81
Find the regression equation and put it on the scatter plot. 81
Evaluate a Formula 81

3.4 EXAMPLE 5 ON THE TI-83 AND TI-83 PLUS 83
Create the scatter plot 83
Find and graph the logarithmic function model for this data 83

3.4 EXAMPLE 5 ON THE TI-86 84
Create the scatter plot 84
Find and graph the logarithmic function model for this data 84

3.4 EXAMPLE 5 USING EXCEL 85
Create the scatter plot 85
Find the regression equation and put it on the scatter plot. 85
Evaluate a Formula 86

CHAPTER 4 HIGHER DEGREE POLNOMIAL AND RATIONAL FUNCTIONS 87

4.1 EXAMPLE 1 ON THE TI-83, TI-83 PLUS, AND TI-84 PLUS 87
Graph the cubic function $y = x^3 - 27x$. 87
Find the local Maximum. 87
Find the local Minimum. 88

4.1 EXAMPLE 1 ON THE TI-86 89
Graph the cubic function $y = x^3 - 27x$ *89*
Find the local Maximum *89*
Find the local Minimum. *90*

4.2 EXAMPLE 2 ON THE TI-83, TI-83 PLUS, AND TI-84 PLUS 91
Create the scatter plot *91*
Find and graph the linear function model for this data *91*
Find and graph the cubic function model for this data *92*

4.2 EXAMPLE 2 ON THE TI-86 93
Create the scatter plot *93*
Find and graph the linear function model for this data *93*
Find and graph the cubic function model for this data *94*

4.2 EXAMPLE 2 USING EXCEL 95
Create the scatter plot *95*
Find the regression equation and put it on the scatter plot. *95*

CHAPTER 5 SYSTEMS OF EQUATIONS AND MATRICES 97

5.2 EXAMPLE 2 ON THE TI-83, TI-83 PLUS, AND TI-84 PLUS 97
Solve the System of Linear Equations with Matrices *97*
Enter the Matrix 97
View the Matrix on the Home Screen 98
Find the reduced row-echelon form of the matrix, rref ([A]) 98

5.2 EXAMPLE 2 ON THE TI-86 99
Solve the System of Linear Equations with Matrices *99*
Enter the Matrix 99
View the Matrix on the Home Screen 100
Find the reduced row-echelon form of the matrix, rref ([A]) 100

5.2 EXAMPLE 2 USING EXCEL 100
Solve the System of Linear Equations with Matrices *100*

5.3 EXAMPLE 2B ON THE TI-83, TI-83 PLUS, AND TI-84 PLUS 101
Perform Matrix Arithmetic *101*
Enter the Matrices as A, B, and C 101
Perform the Matrix Arithmetic on the Home Screen 101
Multiply a matrix by a number 101

5.3 EXAMPLE 2B ON THE TI-86 102
Perform Matrix Arithmetic *102*
Enter the Matrices as A, B, and C 102
Perform the Matrix Arithmetic on the Home Screen 103
Multiply a matrix by a number 103

5.3 EXAMPLE 2B USING EXCEL 104
Perform Matrix Arithmetic *104*
Enter the Matrices as A, B, and C 104
Multiply a matrix by a number 105

5.3 EXAMPLE 6 ON THE TI-83, TI-83 PLUS, AND TI-84 PLUS 106
Perform Matrix Multiplication *106*
Enter the Matrices 106
Display a matrix on the Home Screen 106
Perform the Matrix Multiplication on the Home Screen 106

5.3 EXAMPLE 6 ON THE TI-86 107
Perform Matrix Multiplication *107*
Enter the Matrices 107
Display a matrix on the Home Screen 107
Perform the Matrix Multiplication 107

5.3 EXAMPLE 6 USING EXCEL 108
Perform Matrix Multiplication *108*
Enter the Matrices as B and A 108

5.4 EXAMPLE 3 ON THE TI-83 AND TI-83 PLUS 110
Find the Inverse of a Square Matrix *110*
Enter the Matrix 110
Find the Matrix Inverse on the Home Screen 110
Show that $AA^{-1} = I$ 110

5.4 EXAMPLE 3 ON THE TI-86 111
Find the Inverse of a Square Matrix *111*
Enter the Matrix 111
Find the Matrix Inverse on the Home Screen 111
Show that $AA^{-1} = I$ 111

5.4 EXAMPLE 3 USING EXCEL 112
Find the Inverse of a Square Matrix *112*
Enter the Matrix as A 112

5.4 EXAMPLE 8 ON THE TI-83, TI-83 PLUS, AND TI-84 PLUS 114
Solve a Matrix Equation *114*
Enter the Matrices A and C 114
Find the product matrix of A^{-1} and C on the Home Screen 114

5.4 EXAMPLE 8 ON THE TI-86 115
Solve a Matrix Equation *115*
Enter the Matrices A and C 115
Find the product matrix of A^{-1} and C on the Home Screen 115

5.4 EXAMPLE 8 USING EXCEL 116
Solve a Matrix Equation *116*
Enter the Matrices A and C 116
Find A Inverse 116
Find the product of A Inverse and C 117

CHAPTER 6 SPECIAL TOPICS 119

6.1 EXAMPLE 1 ON THE TI-83, TI-83 PLUS, AND TI-84 PLUS 119
Solve the linear inequality $6x-3y\le 15$ *by graphing.* *119*
Select solid or dotted line using MODE 119
Enter the function 119
Shade the solution region 119

6.1 EXAMPLE 1 ON THE TI-86 120
Solve the linear inequality $6x-3y\le 15$ *by graphing.* *120*
Select solid or dotted line using MODE 120
Enter the function 120
Shade the solution region 120

6.1 EXAMPLE 5 ON THE TI-83 AND TI-83 PLUS 121
Solve the system of linear inequalities $\begin{array}{l} y\le 71-x \\ y\le \dfrac{1260-15x}{28} \\ x\ge 0 \\ y\ge 0 \end{array}$ *by graphing.* *121*
Select solid or dotted line using MODE 121
Enter the functions 121
Shade the solution region 122
Find the corner points *122*

6.1 EXAMPLE 5 ON THE TI-86 123
Solve the system of linear inequalities $\begin{array}{l} y\le 71-x \\ y\le \dfrac{1260-15x}{28} \\ x\ge 0 \\ y\ge 0 \end{array}$ *by graphing.* *123*
Select solid or dotted line using MODE 123
Shade the solution region 124
Find the corner points *124*

6.1 EXAMPLE 5 ON THE TI-84 PLUS 125
Graph a System Linear Inequalities on the TI-84 Plus Silver using Inequalz Apps *125*
Solve the system of linear inequalities $\begin{array}{l} y\le 71-x \\ y\le \dfrac{1260-15x}{28} \\ x\ge 0 \\ y\ge 0 \end{array}$ *by graphing.* *125*
To Quit Inequalz 126

6.3 EXAMPLE 1 ON THE TI-83, TI-83 PLUS, AND TI-84 PLUS 127
Sequence Mode *127*
Calculate the Terms of the Sequence *127*
Graph the sequence $a_n = \frac{8}{n}$ *128*
Set the Window for the Graph of the Sequence *128*
Create a Table of Values for the sequence *129*
Create the graph and table for $b_n = (-1)^n n(n+1)$ *129*

6.3 EXAMPLE 1 ON THE TI-86 130
Sequence Mode *130*
Calculate the Terms of the Sequence *130*
Graph the sequence $a_n = \frac{8}{n}$ *131*
Create a Table of Values for the sequence *131*
Create the graph and table for $b_n = (-1)^n n(n+1)$ *131*

PREFACE

This graphing calculator / EXCEL manual is designed to support the technology examples from the textbook *College Algebra in Context with Applications for the Managerial, Life and Social Sciences, Second Edition* by Ronald J. Harshbarger & Lisa S. Yocco. The technologies covered include Microsoft EXCEL 2000 and the Texas Instruments TI-83, TI-83 PLUS, TI-84 PLUS and TI-86 graphing calculators. The usage directions for the TI-83, TI-83 PLUS, and the TI-84 PLUS are the same, and were created with the TI-84 PLUS Silver Edition.

The manual covers all four technologies for most of the technology examples in this text. The chapters refer to the chapters in the text, and each example includes directions for each technology that is applicable. I encourage users of the graphing calculators to go through the Getting Started section for that calculator before doing other examples.

This manual was created using the TI-83 PLUS, TI-86, EXCEL 2000, and the TI-Graph Link software.

I think that these technologies enhance the study of mathematics at the College Algebra level and hope that this manual will help your understanding of this course. Please feel free to contact me if you have any questions about these technologies for this material.

Florence Chambers
Southern Maine Community College
South Portland, Maine 04106

fchambers@smccme.edu

CHAPTER 1

FUNCTIONS, GRAPHS, AND MODELS

Graphing with Technology - TI-83, TI-83 Plus, and TI-84 PLUS

Getting started

Graphing on the TI-83, TI-83 Plus, and TI-84 Plus is handled in the same way. There are five graphing keys just below the screen: [Y=], [WINDOW], [ZOOM], [TRACE], and [GRAPH]. You may input up to 10 functions in the [Y=] menu. Each function is identified by number, Y1=, Y2=, ..., Y9=, Y0=.You may also graph a set of data points using Plot 1, Plot 2, or Plot 3 found on the graph screen above the Y= functions. Let's go through the entire process of graphing functions using $y=-x^2+2x-1$ and $y=x^3-3x^2-13$.

Preliminaries to graphing on the TI-83, TI-83 Plus, and TI-84 PLUS

- The function *must* be solved for *y*.
- Use only the variable *x* in the equation.
- Input *x* with the [X,T,Θ,*n*] key.
- Use the subtraction key [-] between terms.
- Use the negation key [(-)] when the first term is negative.
- The negation key can be found in the last row of keys near the ENTER key.

Input the function $y=-x^2+2x-1$

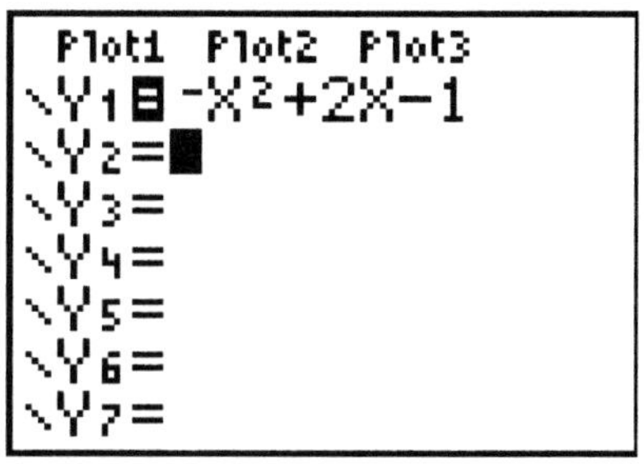

- Press the [Y=] key below the screen.
- Position the cursor on \Y1=.
- Press [(-)] [X,T,Θ,*n*] [x^2] [+] [2] [X,T,Θ,*n*] [-] [1].
- The = sign is now highlighted to show that Y1 is *on* and ready to be graphed.

Window Settings

- Press the [WINDOW] key below the screen.
- The current settings of your calculator window are shown.
- You may change the settings by hand or use the [ZOOM] key to automatically set special windows, such as the Standard Window [-10, 10, 1] for [Xmin, Xmax, Xscl] by [-10, 10, 1] for [Ymin, Ymax, Yscl] described next.

Setting the Standard Window [-10, 10, 1] by [-10, 10, 1]

- Many functions can be graphed in the standard window. You may enter these values by hand or use the Zoom Standard feature of your calculator as follows.
- Press the [ZOOM] key.

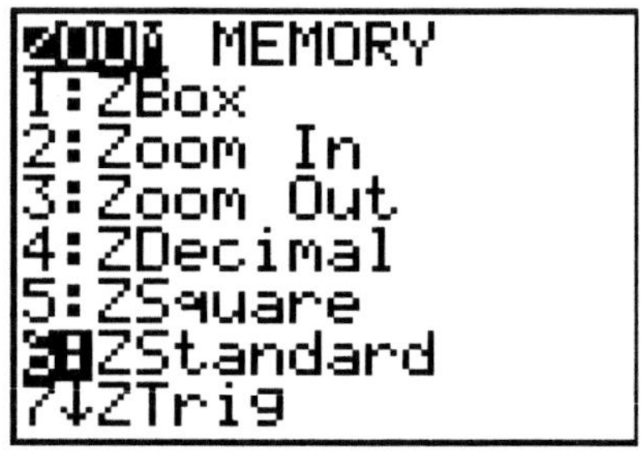

- Select 6:Zstandard. The function(s) that are *on* are automatically redrawn in the standard window.
- Press the [WINDOW] key below the screen to see the Standard Window settings.

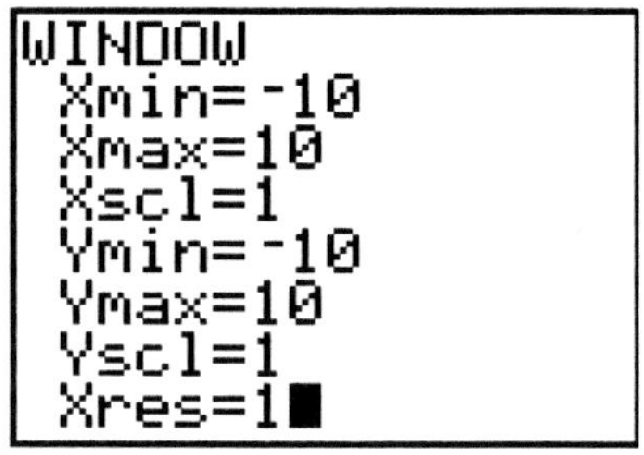

- The standard window is very appropriate for Y1 = $-x^2+2x-1$ as shown below.

Graph the function(s)

- Press the [GRAPH] key below the screen.
- You will see the graph of any function that is turned *on*.

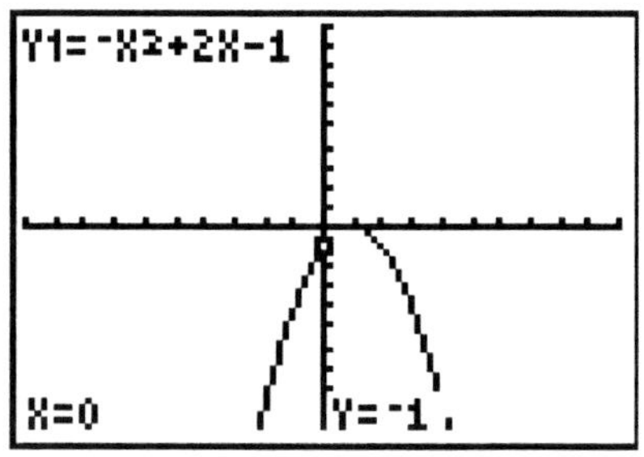

Trace the function

- Press the [TRACE] key below the screen.
- Use the left and right arrow keys to walk along the graph.
- The bottom of the screen displays the *x* and *y* coordinates of the point that the cursor is on.
- If more than one function is graphed on the same screen, you can use the up and down arrows to move from one function to another.

Display the equation on the graph

- The graph screen can be formatted by using the FORMAT feature.
- FORMAT is found above the [ZOOM] key. In particular, you can display the equation.
- Press [2nd] [ZOOM].
- Use the arrow keys to place the cursor over the last line, ExprOn.
- I suggest that the settings shown on the right be used.

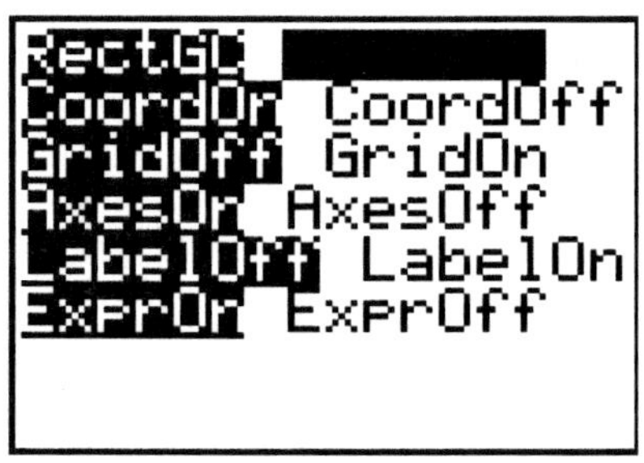

- Press [ENTER].
- Whenever you use [TRACE] on a graph, the equation will be displayed.

Input the function $y = x^3 - 3x^2 - 13$.

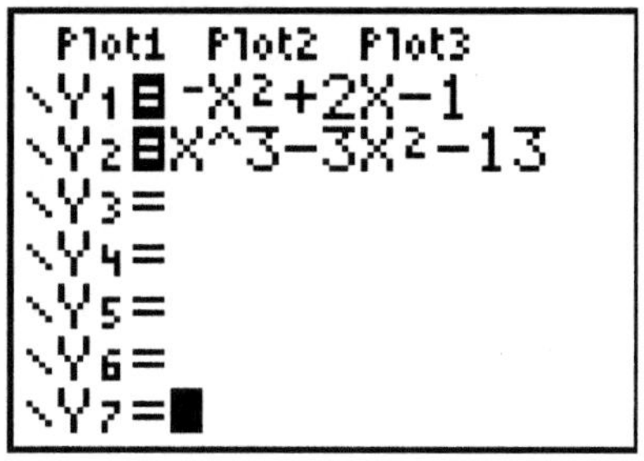

- Press the [Y=] key below the screen.
- Position the cursor on \Y2=.
- Press [X,T,Θ,n] [^] [3] [−] [3] [X,T,Θ,n] [x²] [−] [1] [3].
- The = sign is now highlighted to show that Y2 is *on* and ready to be graphed.

Turn functions on and off

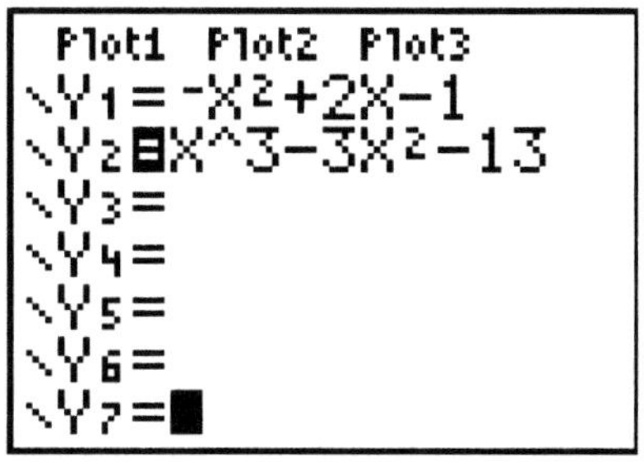

- A function must be *on* to be graphed.
- A function is *on* when the equal sign is highlighted.
- To toggle between *on* and *off*, use the arrow keys to position the cursor over the equal sign. Press [ENTER]. The function will now be *on* if was *off*, and *off* if it was *on*.
- Turn off Y1.

Set the viewing window for Y2 = $x^3 - 3x^2 - 13$.

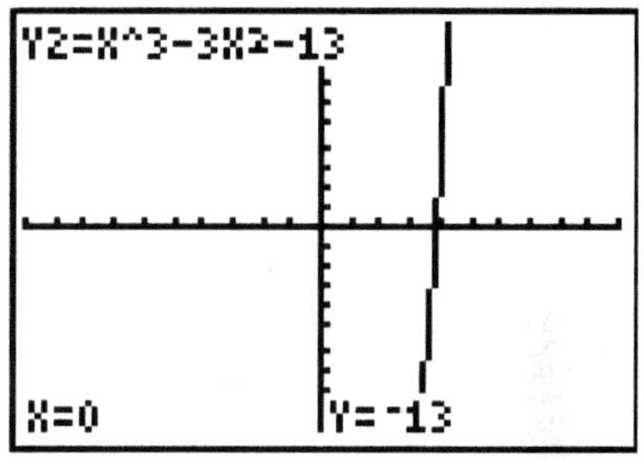

- The window setting defines the highest and lowest values of x and y on the graph of the function that will be shown on the screen. The window should show all the important parts of a graph.
- The window is currently set to the standard viewing window. Note that this window is not appropriate for Y2 = $x^3 - 3x^2 - 13$ because this window does not show the turning point of the graph.

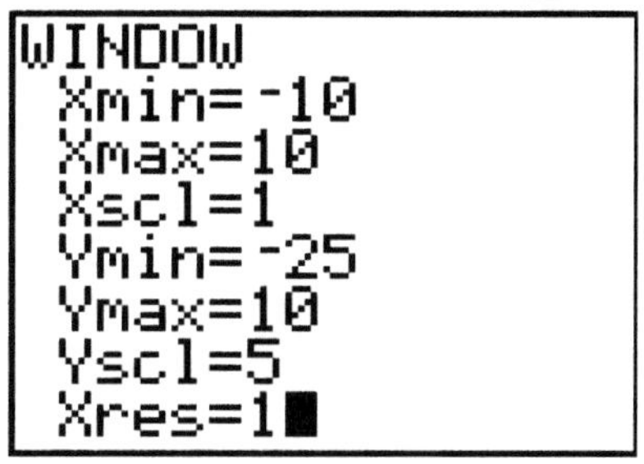

- Press the [WINDOW] key below the screen.
- Use the arrow keys to position the cursor on the line you wish to change.
- Set the window values as shown.
- Use Yscl = 5 so that there are not so many tick marks on the y-axis.

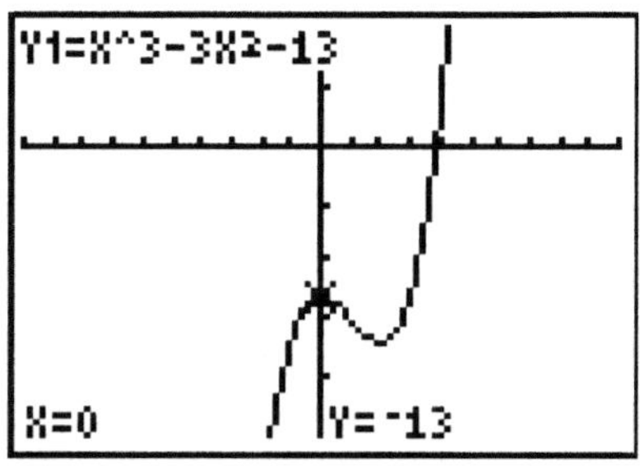

When more than the Graph is displayed

- If Plot 1, Plot 2, or Plot 3 is highlighted on the Y= screen, you may see more than the graph of your function, perhaps a histogram or a set of dots.
- From the Y= screen, turn off the Plot.
- Use the up arrow key to position the cursor over the Plot that is turned on.
- Press [ENTER] to turn it off.

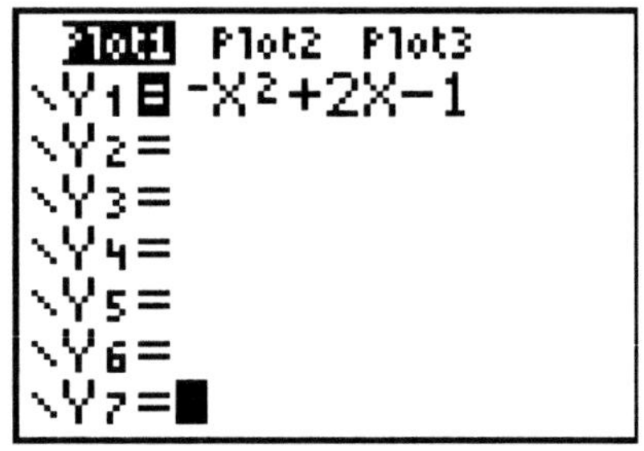

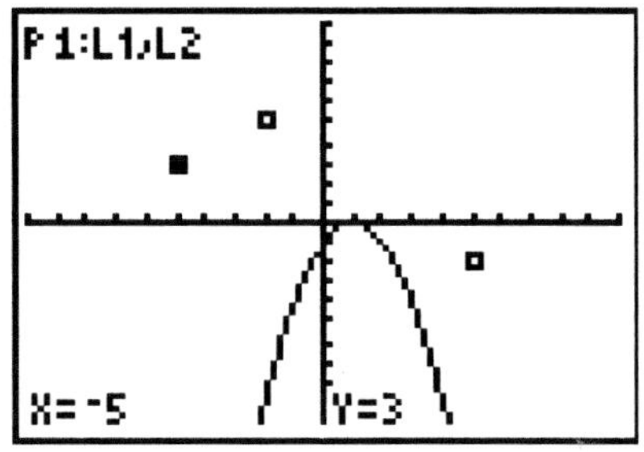

When you get an Invalid Dim Error message

- If Plot 1, Plot 2, or Plot 3 is highlighted on the Y= screen, you may also get a dimension error and no Y= equations can be graphed.
- Invalid Dim error means that there is no data in the lists defined in the highlighted Plot.
- To be able to graph your equations, turn off the plot.
- Use the up arrow key to position the cursor over the Plot that is turned on.
- Press [ENTER] to turn it off.
- All Y= equations that are turned on will now graph.

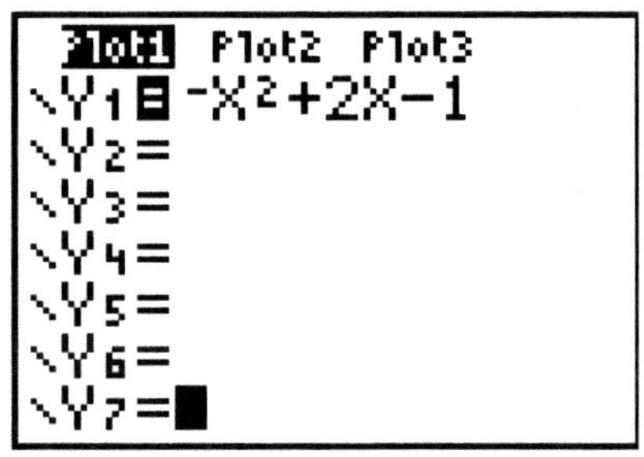

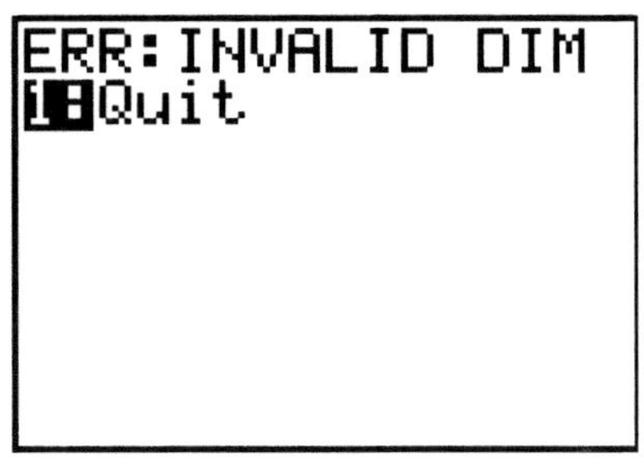

Graphing with Technology - TI- 86

Getting started

Graphing on the TI-86 is similar to graphing on the TI-83 and the TI-83 Plus. The TI-86 uses a menu system on the graph screen to identify the graphing features. The keys below the graph screen are labeled as F1, F2, F3, F4, and F5. You must press the GRAPH key located below the blue ALPHA key to access the graph menu. You may input up to 99 functions in the y(x)= menu. Each function is identified by number, y1=, y2=, … , y99=.

You may also graph a set of data points using Plot 1, Plot 2, or Plot 3 found on the graph screen above the y = functions. Let's go through the entire process of graphing functions using $y=-x^2+2x-1$ and $y=x^3-3x^2-13$.

Preliminaries to graphing on the TI-86

- The function *must* be solved for *y*.
- Use only the variable *x* in the equation.
- Input *x* with the x-VAR key.
- Use the subtraction key − between terms.
- Use the negation key (-) when the first term is negative.

Input the function $y=-x^2+2x-1$

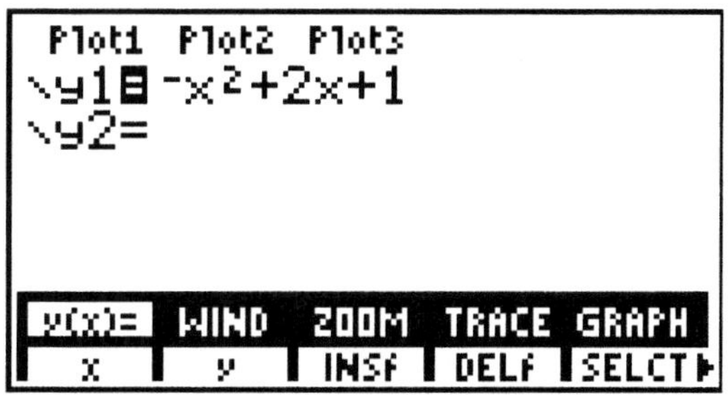

- Press the GRAPH key on the keyboard. It is below the blue ALPHA key.
- Press the F1 key to access the y(x)= menu the screen.
- Position the cursor on \y1=.
- Press (-) x-VAR x^2 + 2 x-VAR − 1.
- The = sign is now highlighted to show that y1 is *on* and ready to be graphed.
- The negation key can be found in the last row of keys near the ENTER key.

Window Settings

- Press the WINDOW key below the screen.
- The current settings of your calculator window are shown.

- You may change the settings by hand or use the [ZOOM] key to automatically set special windows, such as the Standard Window [-10, 10, 1] for [Xmin, Xmax, Xscl] by [-10, 10, 1] for [Ymin, Ymax, Yscl] described next.

Setting the Standard Window [-10, 10, 1] by [-10, 10, 1]

Many functions can be graphed in the standard window. You may enter these values by hand or use the Zoom Standard feature of your calculator as follows.

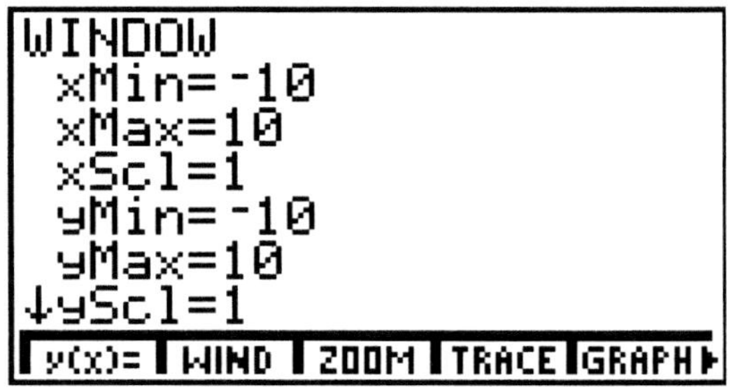

- Press the [F3] key below the screen to access the ZOOM menu.
- Select [F4] ZSTD. The function(s) that are *on* are automatically redrawn in the standard window.
- Press [EXIT] to remove the second menu from the graph screen.
- The standard window is very appropriate for y1 = $-x^2+2x-1$ as shown below.

Graph the function(s)

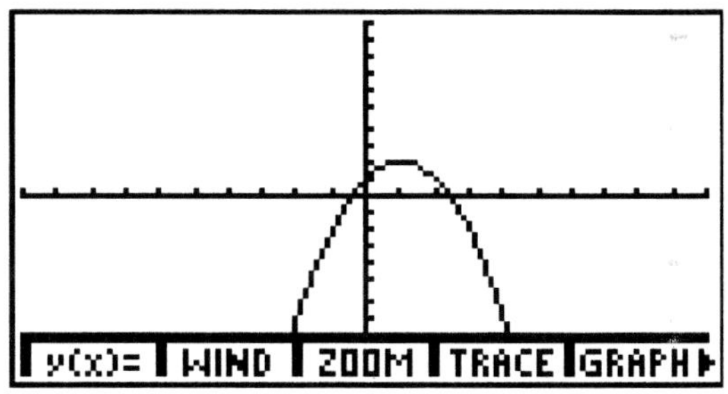

- You will see the graph of any function that is turned *on*.
- Press [2nd] [F5] or [EXIT] [F5]to view the graph of Y1.
- Press the [F5] key to graph y1.
- Press [CLEAR] to remove the menu line from the graph screen.

Trace the function

- Press the [F4] key below the graph screen to use TRACE.
- Use the left and right arrow keys to walk along the graph.

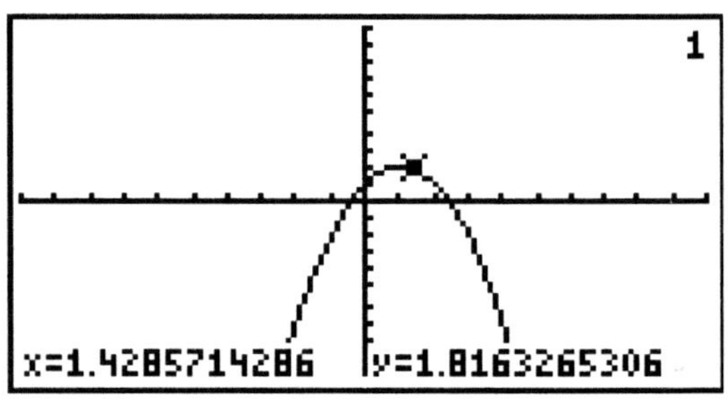

- The bottom of the screen displays the x and y coordinates of the point that the cursor is on.
- The number 1 in the upper right corner of the screen tells you that you are looking at the graph of the function defined as y1.
- If more than one function is graphed on the same screen, you can use the up and down arrows to move from one function to another.
- Press [EXIT] to return to the graph menu on the screen.

Display the equation on the graph

- The TI-86 does not have the feature that can display the equation of y= functions on the graph.
- The equation or plot is identified by 1, 2, 3,..., or P1, P2, P3 in the upper right corner.

Input the function $y = x^3 - 3x^2 - 13$.

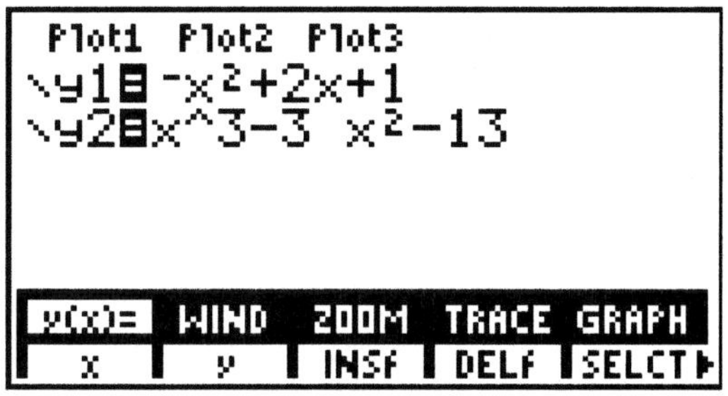

- Press the [GRAPH] key on the keyboard. It is below the blue [ALPHA] key.
- Press the [F1] key to access the y(x)= menu the screen.
- Press the down arrow to go to the next line.
- Position the cursor on \y2=.
- Press [x-VAR] [^] [3] [−] [3] [x-VAR] [x^2] [−] [1] [3].
- The = sign is now highlighted to show that you y2 is *on* and ready to be graphed.

Turn functions on and off

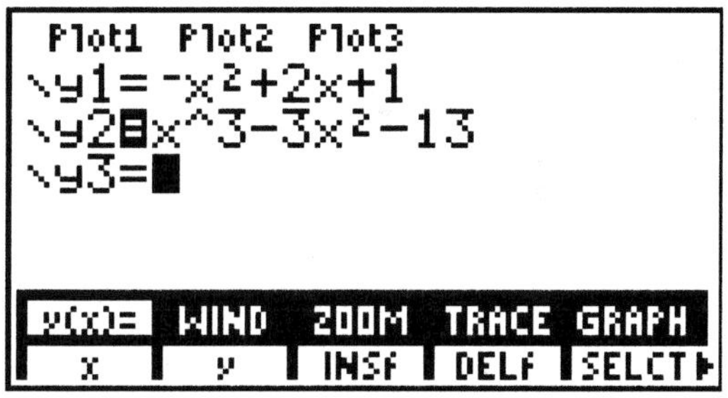

- A function must be *on* to be graphed.
- A function is *on* when the equal sign is highlighted.
- To toggle between *on* and *off*, you must be in the y(x)= menu and use SELCT.
- Press the [F1] key under the y(x)= menu . Another menu appears below the y(x)= menu.
- Position the cursor on the y1= function.
- Press the [F5] key to use SELCT. This turns y1 on or off.
- Press the [EXIT] key to see only one line of menu again.

Set the viewing window for y2 = $x^3 - 3x^2 - 13$.

The window setting defines the highest and lowest values of *x* and *y* on the graph of the function that will be shown on the screen. The window should show all the important parts of a graph.

- The window setting defines the highest and lowest values of *x* and *y* on the graph of the function that will be shown on the screen. The window should show all the important parts of a graph.
- The window is currently set to the standard viewing window. Note that this window is not appropriate for Y2 = $x^3 - 3x^2 - 13$ because this window does not show the turning point of the graph.

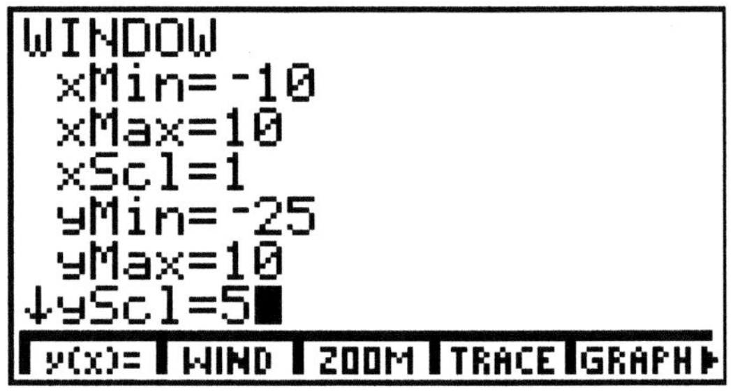

- If one menu line is shown on the graph screen, press the [F2] key below the screen to select WIND.
- If two menu lines are shown, press [F2] [EXIT] or [2nd] [F2] to select WIND.
- Use the arrow keys to position the cursor on the line you wish to change.
- Enter the values that you want. Set the values as shown.
- Use yScl = 5 so that there are not so many tick marks on the y-axis.

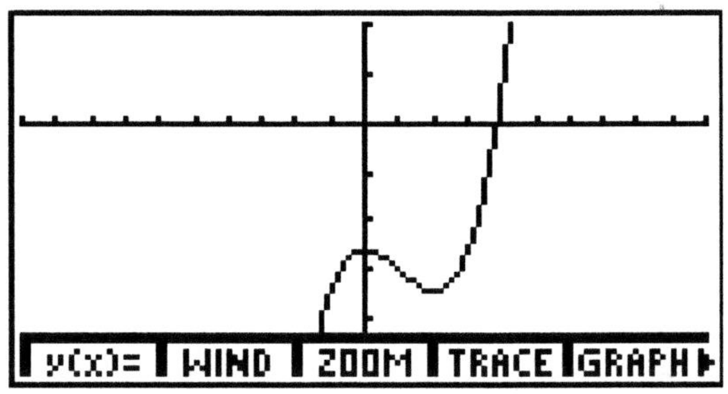

- Press [F5] GRAPH to see the graph of y2.
- The number 2 in the upper right hand corner of the screen tells you that you are looking at the graph of the function defined as y2.
- Press [F4] to Trace the function.

When more than the Graph is displayed

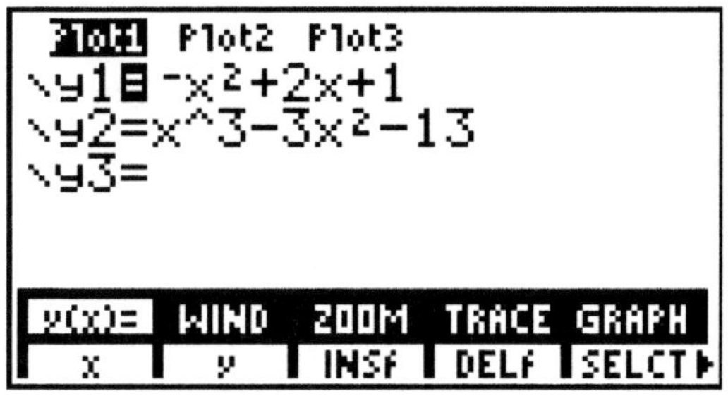

- If Plot 1, Plot 2, or Plot 3 is highlighted on the Y= screen, you may see more than the graph of your function, perhaps a histogram or a set of dots.
- From the Y= screen, turn off the Plot.
- Use the up arrow key to position the cursor over the Plot that is turned on.
- Press [ENTER] to turn it off.

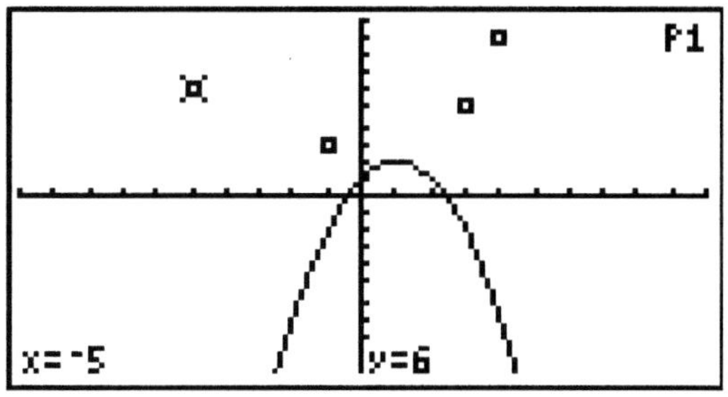

When you get an Invalid Dim Error message

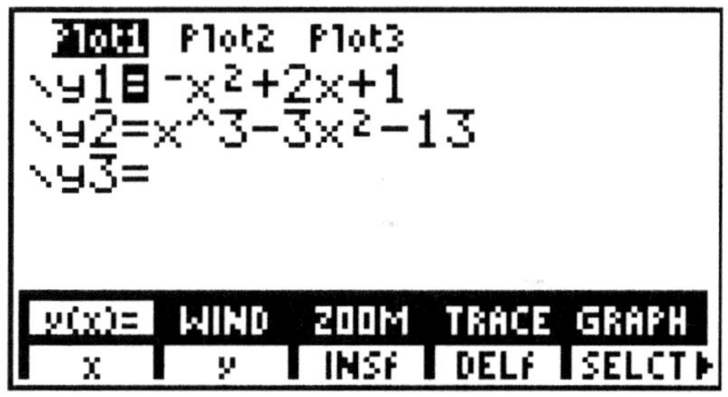

- If Plot 1, Plot 2, or Plot 3 is highlighted on the Y= screen, you may also get a dimension error and no Y= equations can be graphed.
- Invalid Dim error means that there is no data in the lists defined in the highlighted Plot.
- To be able to graph your equations, turn off the plot.
- Use the up arrow key to position the cursor over the Plot that is turned on.
- Press [ENTER] to turn it off.
- All Y= equations that are turned on will now graph.

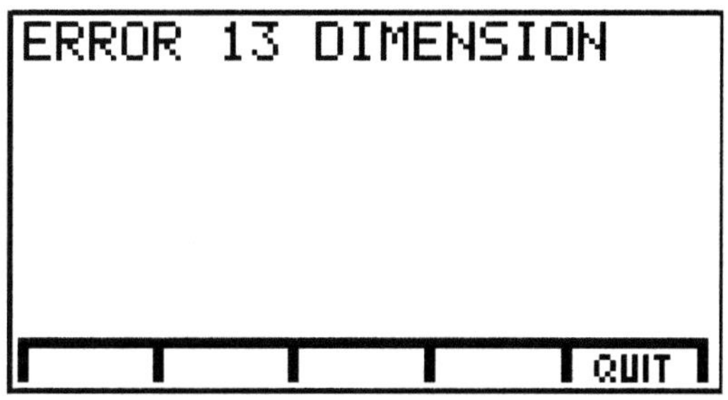

1.2 Example 2 on the TI-83, TI-83 PLUS and TI-84 PLUS

Graph the Function $C = \dfrac{5350p}{100-p}$

In Example 2 you are asked to graph the function $C = \dfrac{5350p}{100-p}$, where C is the cost in dollars to remove p percent of pollution from drinking water.

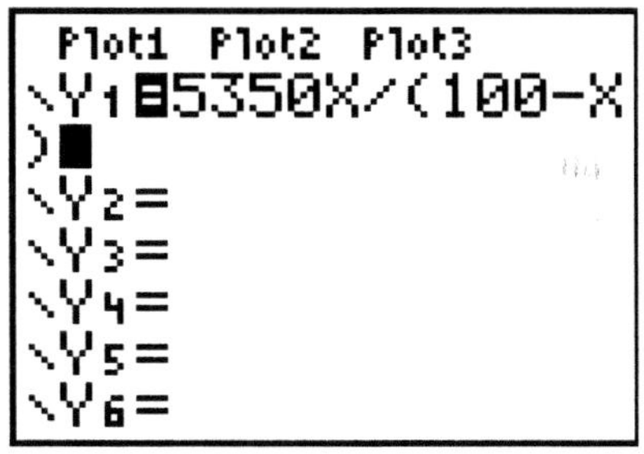

Enter the Equation

Press the [Y=] key to enter the function as $Y1 = 5350x/(100-x)$. Note that the parentheses in the denominator are *essential.*

Set the Window

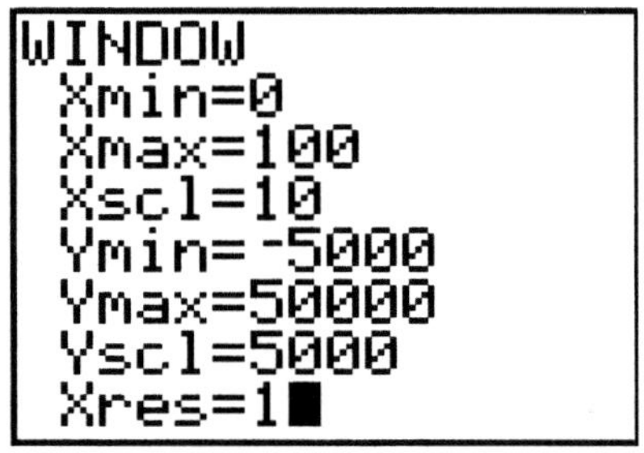

- Use the [WINDOW] key to set the window as shown.
- Set Xscl=10 and Yscl=5000.
- Set Ymin= - 5000 so that you can see the bottom of the graph.
- Set Ymax= 50000.

Graph the equation

- Press the [GRAPH] key to see the graph.

Find *x* = 90 directly on the graph

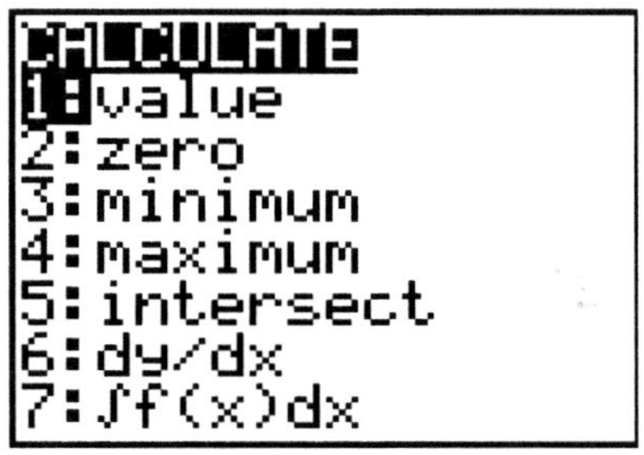

One way to find x=90 on the graph is to use CALC.

- Press [2nd] [TRACE].
- Press the [1] key to select 1:value. The graph appears on the screen.
- Enter [9] [0] [ENTER] at the blinking cursor to see both coordinates.

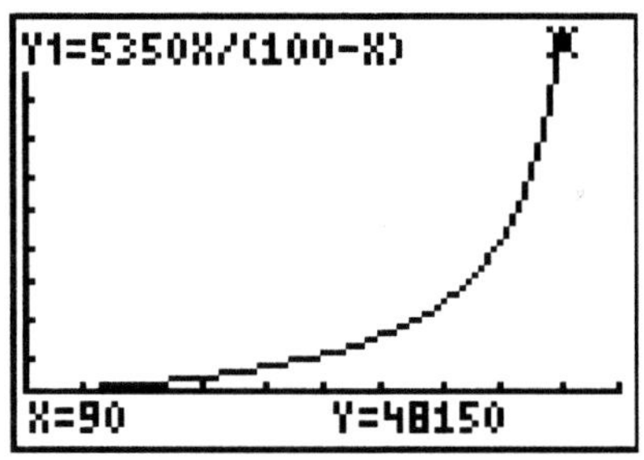

Another way to find x=90 on the graph is to use [TRACE].

- Graph the function.
- Press [TRACE].
- Enter the value 90 at the bottom of the screen.
- Press [ENTER] to see the resulting y value, 48150.

Find *x* = 90 using the TABLE ASK feature

- Enter the function and set the window as described above.
- To make a table, press [2nd] [WINDOW] to set the table using TBLSET.
- Start the table at 0 and select Indpnt: Ask.

- Press [2nd] [GRAPH] to see the table. Enter [9] [0] [ENTER] at the bottom of the screen. Note the *y*-value 48150.

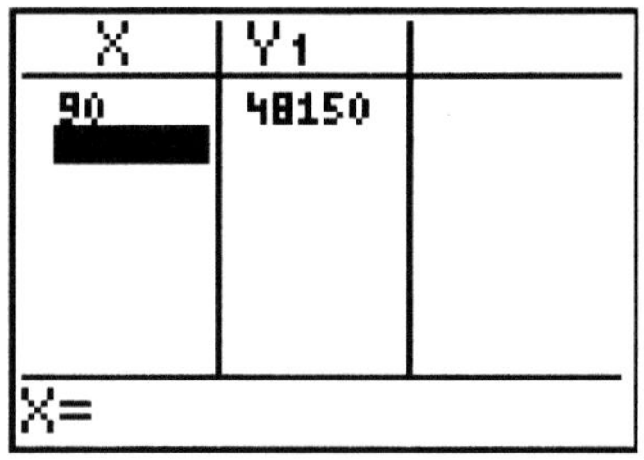

Find *x* = 90 by making a TABLE of values

- Enter the function and set the window as described above
- To make a table, press [2nd] [WINDOW] to set the table using TBLSET.
- Start the table at 0.
- Set Δ Tbl = 15.
- Select Indpnt: Auto.
- Press [2nd] [GRAPH] to see the table. Locate *x* =90 in the table. Note the *y*-value 48150.

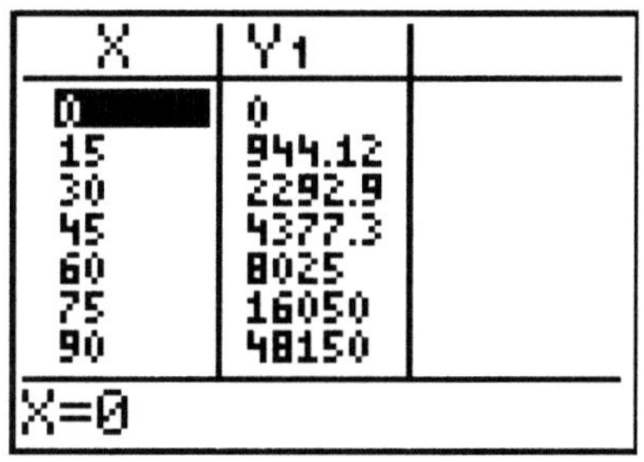

1.2 Example 2 on the TI-86

Graph the Function $C = \dfrac{5350p}{100-p}$ on the TI-86

In Example 2 you are asked to graph the function $C = \dfrac{5350p}{100-p}$, where C is the cost in dollars to remove p percent of pollution from drinking water.

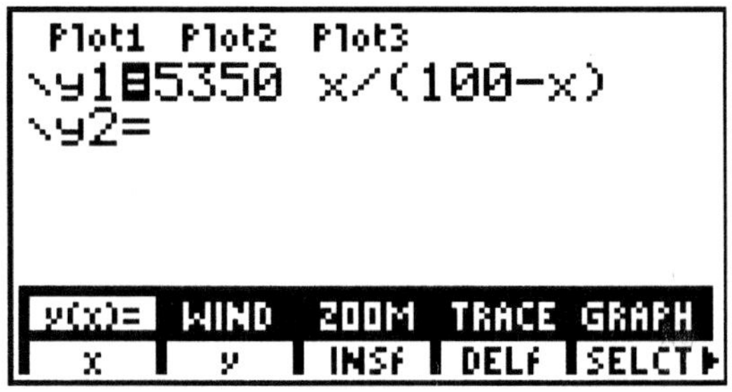

Enter the Equation

- Press the [GRAPH] key, then the [F1] key to enter the function in the y(x)= menu as $y1 = 5350x/(100-x)$. Note that the parentheses in the denominator are *essential.*

Set the Window

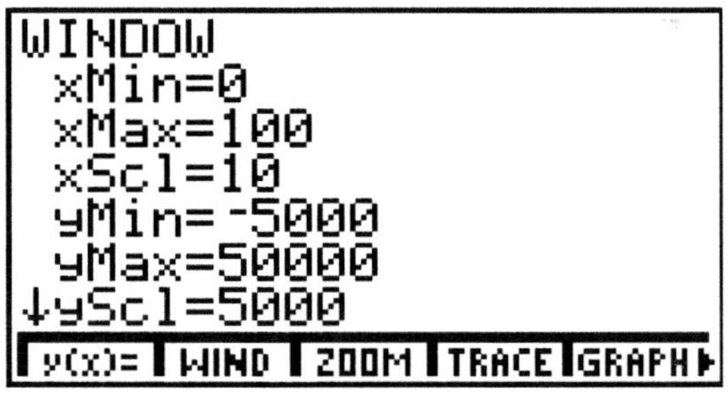

- If two menu lines are showing on the screen, press the [EXIT] key once to see only one line.
- Press the [F2] key WIND to set the window as shown.
- Note that you may also access the Window screen by pressing [2nd] [F2] when two menu lines are showing on the screen.
- Set xScl=10 and yScl=5000.
- Set yMin= - 5000 so that you can see the bottom of the graph.
- Set yMax= 50000.

Graph the equation

- Press [F5] GRAPH to see the graph of the function.
- Use [F4] TRACE to walk along the function.

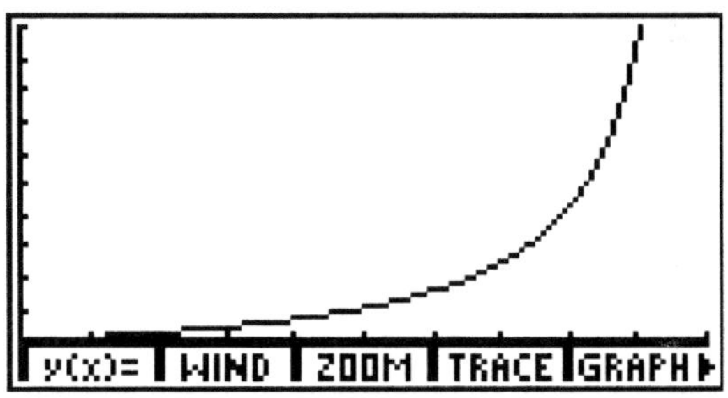

Find *x* = 90 directly on the graph

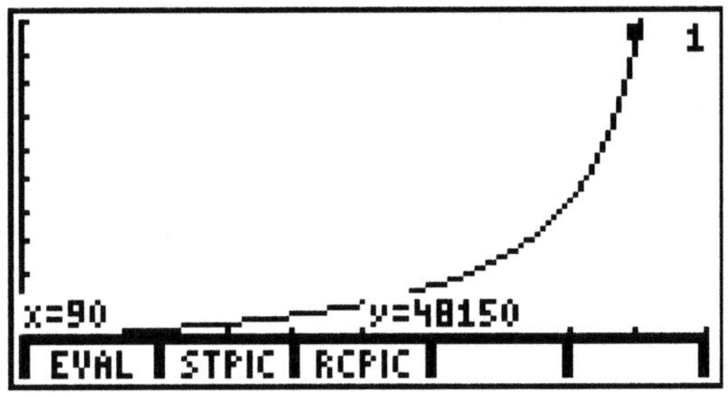

One way to find *x*=90 on the graph is to use EVAL in the Graph menu.

- Press [F5] to see the graph.
- Press [MORE] [MORE] to locate the EVAL function.
- Press the [F1] key to select EVAL.
- Enter [9] [0] [ENTER]at the blinking cursor to see both coordinates.

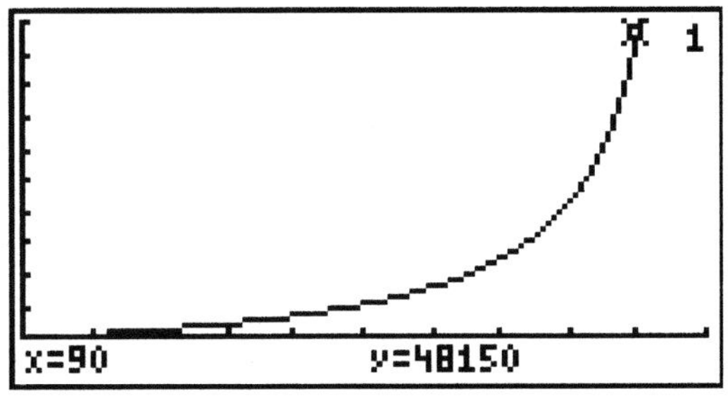

Another way to find *x*=90 on the graph is to use TRACE..

- Graph the function.
- Press [F4] TRACE.
- Enter the value 90 at the bottom of the screen.
- Press [ENTER] to see the resulting y value, 48150.

Find *x* = 90 using the TABLE ASK feature

- Enter the function in the y(x)= menu and set the window as described above.
- To make a table, press the [TABLE] key to access the table menu.
- Press [F2] to set the table using TBLSET.
- Start the table at 0 and select Indpnt: Ask.
- Press [F1] TABLE to see the table. Enter [9] [0] [ENTER] at the bottom of the screen. Note the *y*-value 48150.

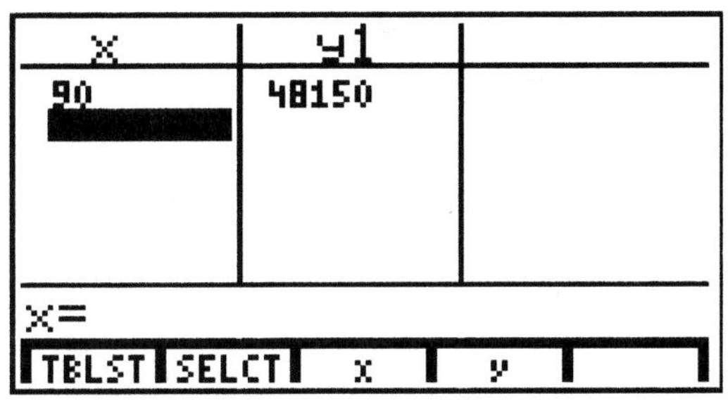

Find *x* = 90 by making a TABLE of values

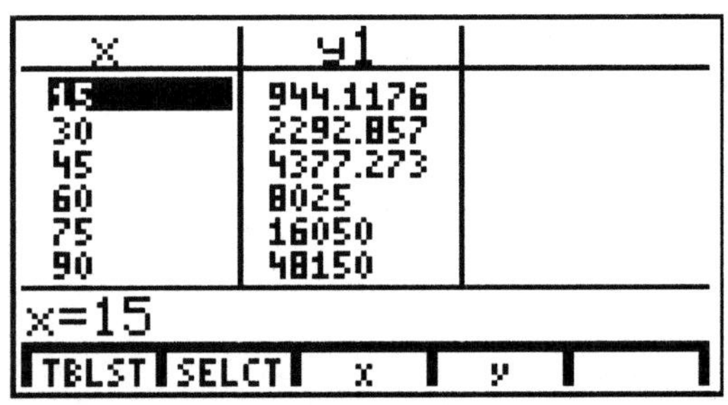

- Enter the function and set the window as described above.
- To make a table, press the [TABLE] key to access the table menu.
- Start the table at 15.
- Set Δ Tbl = 15.
- Select Indpnt: Auto.
- Press [F1] TABLE to see the table. Locate *x* =90 in the table. Note the *y*-value 48150.

1.2 Example 2 using EXCEL

Graph the Function $C = \dfrac{5350p}{100-p}$ in EXCEL

In Example 2 you are asked to graph the function $C = \dfrac{5350p}{100-p}$, where C is the cost in dollars to remove p percent of pollution from drinking water.

To graph a function in EXCEL, you must first enter the data set, then use X-Y Scatter in the Chart Wizard to create the graph of the function.

Enter the Data Set

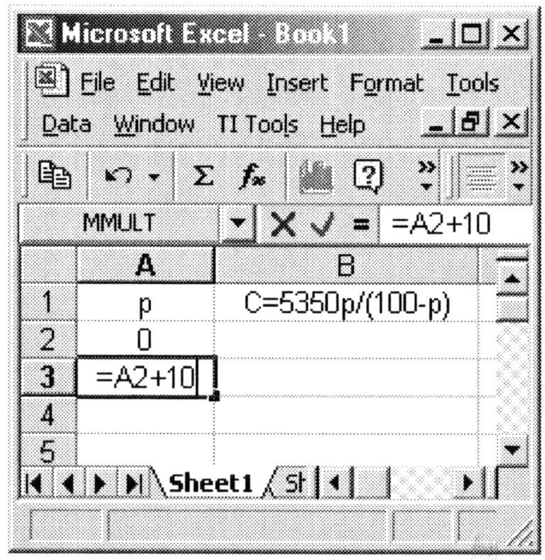

- Open a new worksheet in EXCEL.
- In cell A1, enter p, and in B1 enter the letter C and the formula for C as shown.
- To create the data set quickly, choose a regular increase in value of p, such as 10.
- In cell A2 enter the first value, 0.
- In cell A3 type the formula **=A2+10**. To do this by referencing the A2cell, type the = sign in A3, then left click on the A2 cell.
- Click back in cell A3 and type +10.
- Press ENTER to see the value 10 in A3.
- Go back to cell A3.
- Use the lower right handle of the A3 cell to drag the formula through cell A11 where p = 90.
- Enter the formula for C in cell B2. Enter **=5350*A2/(100-A2)**. Again, reference cell A2 by clicking in it.

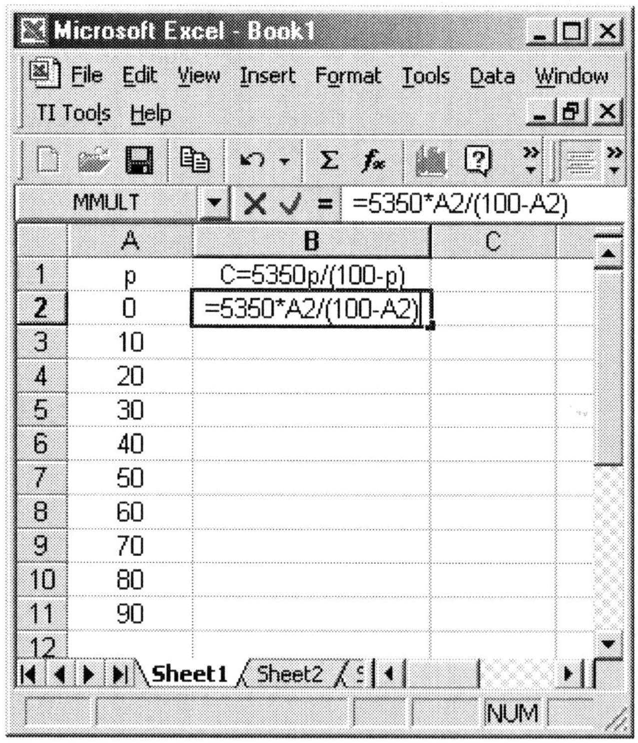

- Press ENTER.
- Use the lower right handle of B2 to drag the formula through cell B11.

Format column B to currency notation

- Go to Format on the toolbar.
- Highlight the values in column B. To do this, go to cell B2 and, holding the left mouse button, drag thru B11.
- Select Format Cells to Currency as shown.
- Click OK.
- Column B is now in currency notation.
- You may also center the values in the column by highlighting the column(s) you wish centered, and then clicking on the center button on the toolbar.

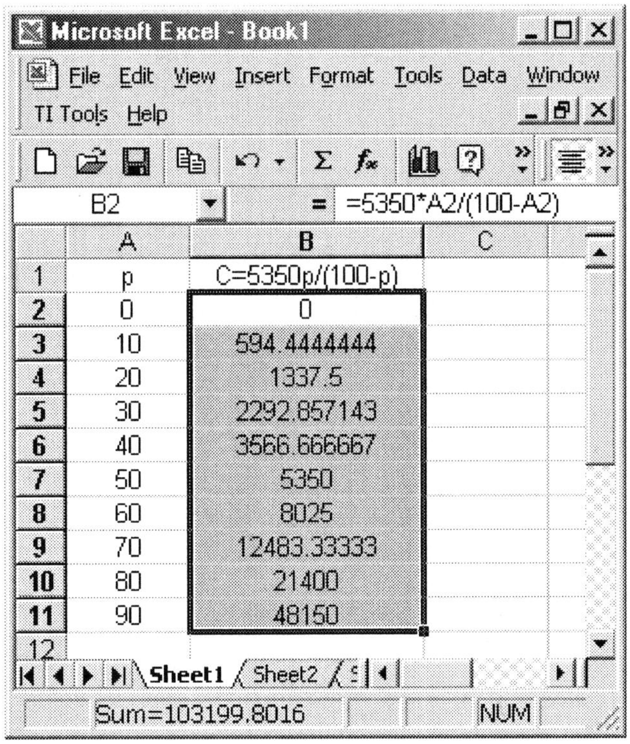

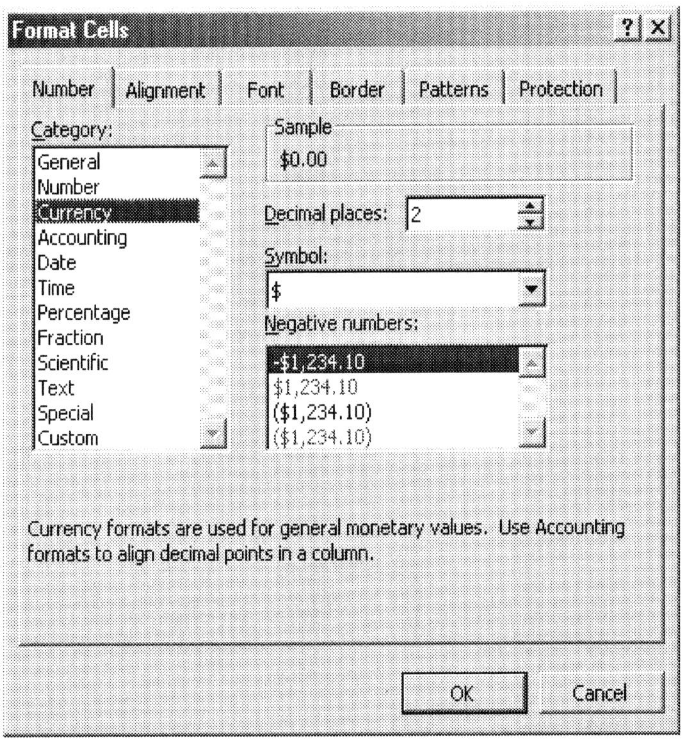

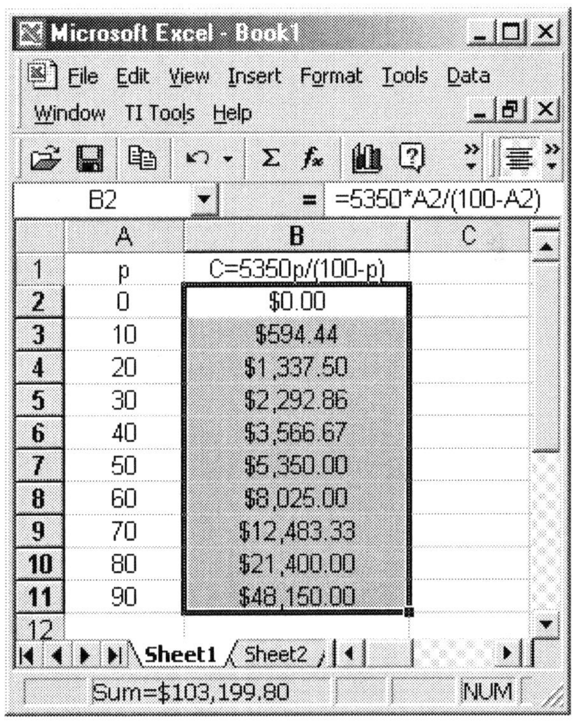

Graph the Equation

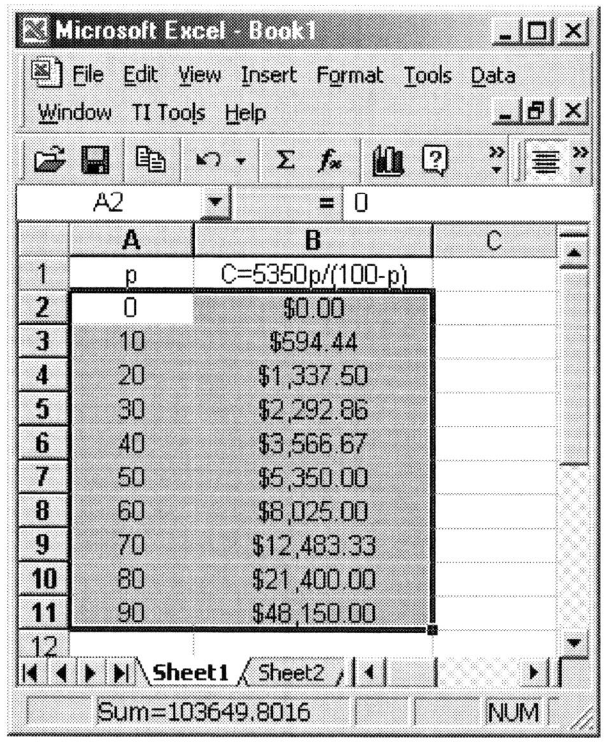

- To graph the function, $C = \dfrac{5350p}{100-p}$, first select the data set. To do this, left click on cell A1 and hold the left mouse button down until you highlight the entire table that contains the data.
- Select **Chart Wizard** on the menu bar.
- Select *XY* Scatter.
- Select Scatter data points connected by lines without markers. This is the last selection of scatter plot in the bottom right corner. You could also use Scatter data points connected by smoothed lines without markers.

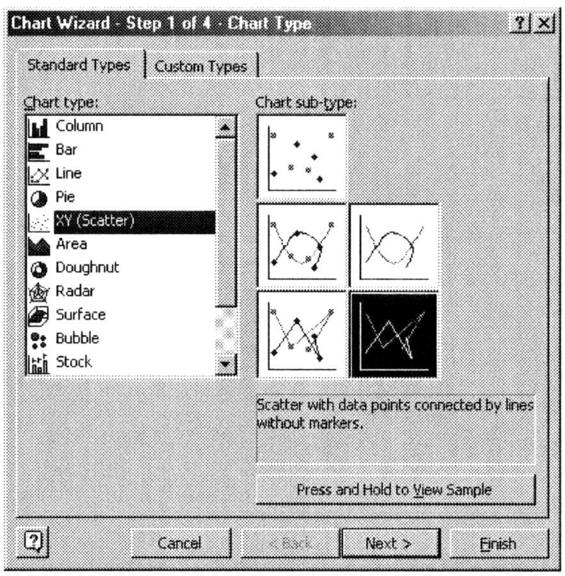

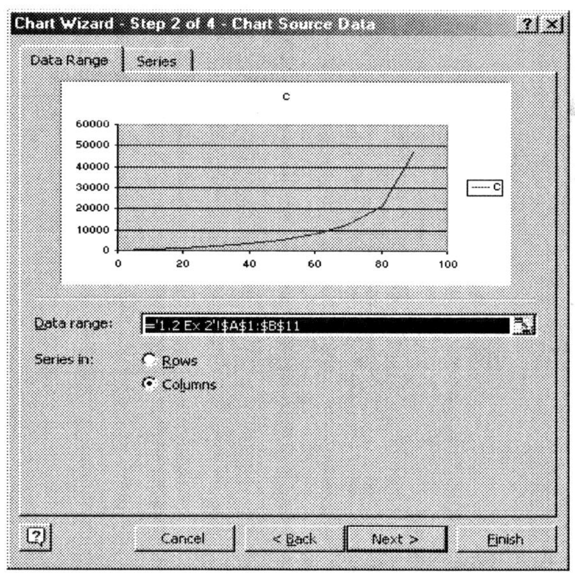

- Click on **NEXT.**
- Data range should be the range of cells you highlighted.
- Check Series in columns.
- Click on **NEXT.**
- Click on the **Titles** tab and fill in the titles for the chart and the x and y axes.
- Click on the **Gridlines** tab and *uncheck* **major gridlines**.

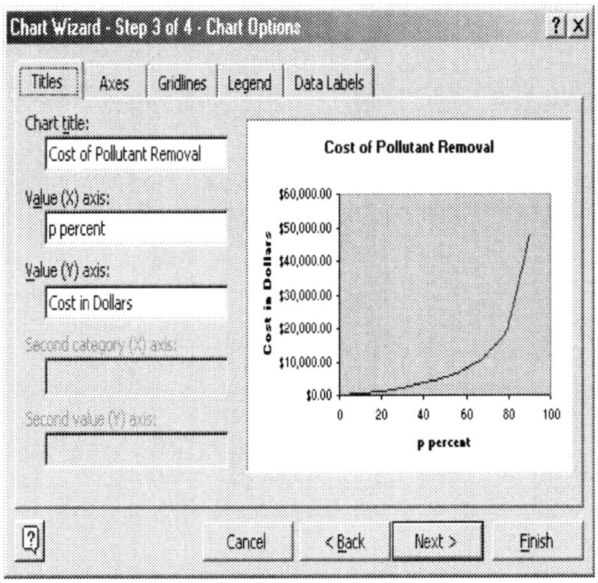

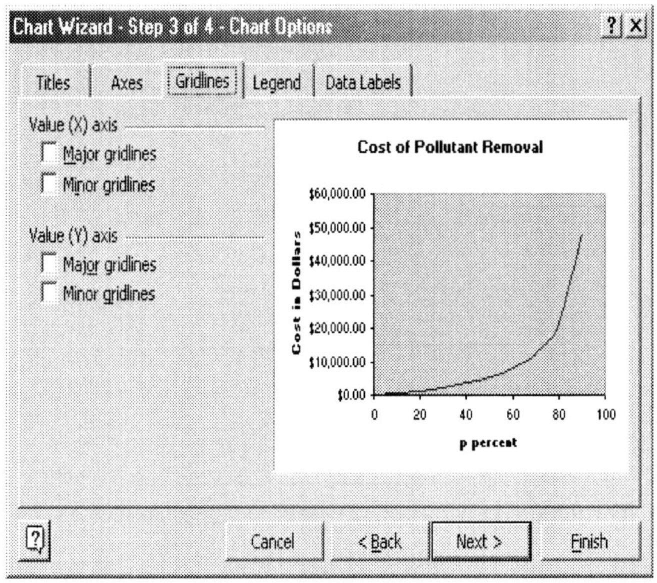

- Click on **Legends** tab and *uncheck* **Show Legend.**
- Click on **NEXT.**
- Select As object in.
- Click on **FINISH.** The scatter plot will appear as a graph in EXCEL.
- You may right click on the chart to format the chart at any time.

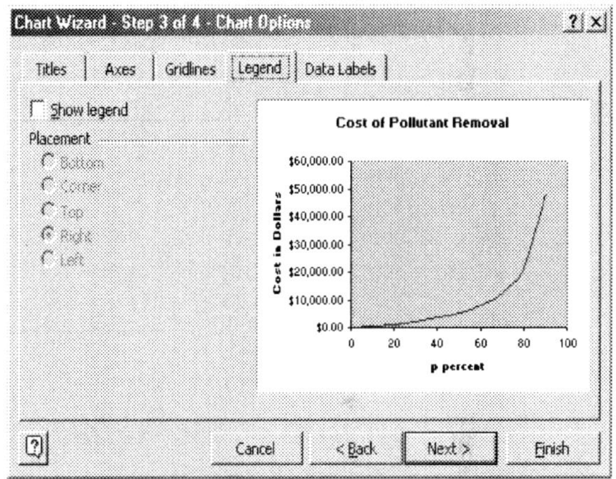

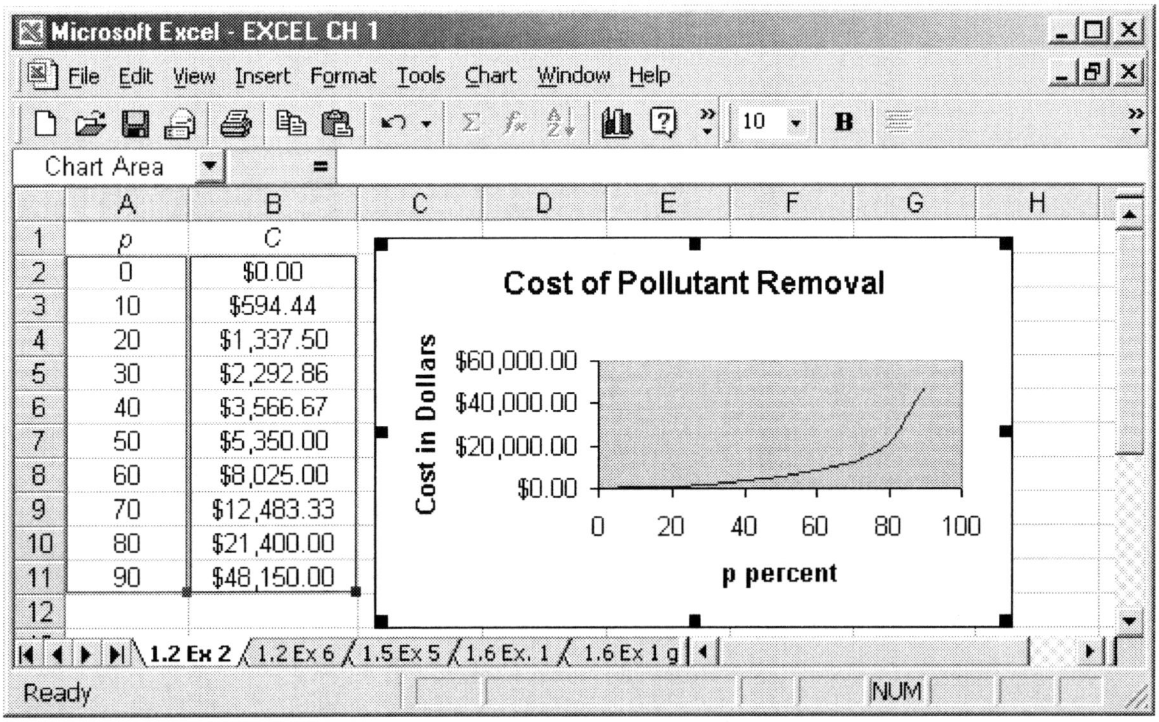

1.2 Example 5 on the TI-83, TI-83 Plus, and TI-84 Plus

Create a Scatter Plot of Data Points on the TI-83, TI-83 Plus, and TI-84 Plus

Example 5 asks you to create a scatter plot of data points. This is handled the same way on both the TI-83, the TI-83 Plus and the TI-84 PLUS. Use the STAT menu of the calculator and define and select Plot 1, Plot 2, or Plot 3.

TABLE 1-12

Decade	Years from 1900 to End of Decade	Number of Executions
1930-1939	40	1690
1940-1949	50	1284
1950-1959	60	717
1960-1969	70	191
1970-1979	80	3
1980-1989	90	117
1990-1999	100	518

Set the Window

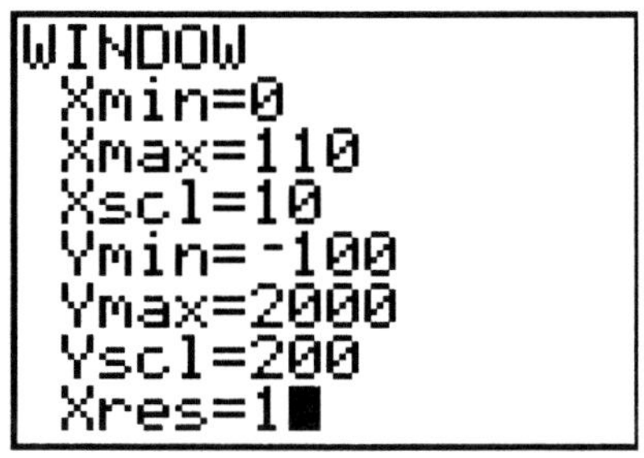

- First set the window.
- Press the [WINDOW] key.
- After you examine the data, choose window values that include these values and some "space" above and below the highest and lowest values of x and y.
- Set the window elements to

  Xmin=0, Xmax=110, Xscl=10, Ymin= - 100, Ymax=2000, Yscl=200.
- Press the down arrow or [ENTER] to move from line to line.

Enter the Data Points

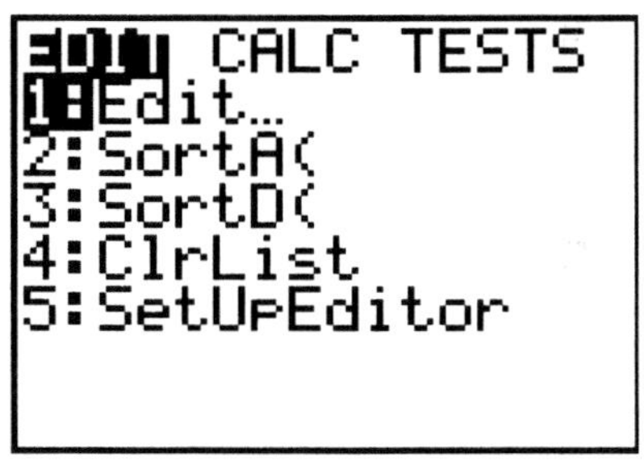

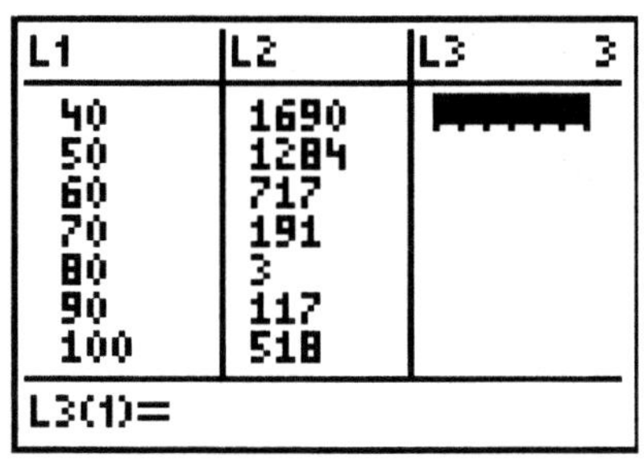

- Press the [STAT] key.
- Press [1] to select 1:Edit or move the cursor to 1:Edit and press [ENTER]. This brings you to the screen where you can enter the data in the lists.
- You may need to clear the lists. See directions below.
- Enter the Years from 1900-values in list L_1.
- Enter the Number of executions-values in L_2.
- Use the down arrow to move from line to line.

Clearing Lists

Clear Lists from the Home screen:

- If you wish to clear the lists, go back to STAT EDIT and select 4: ClrList. This pastes ClrList on the Home Screen.
- Press [2nd] [1] (L_1) [ENTER] to clear L_1. You should see the word Done.
- Repeat the line by pressing [2nd] [ENTER] (ENTRY), and move the cursor over L_1. Press [2nd] [2] (L_2) to clear L_2.

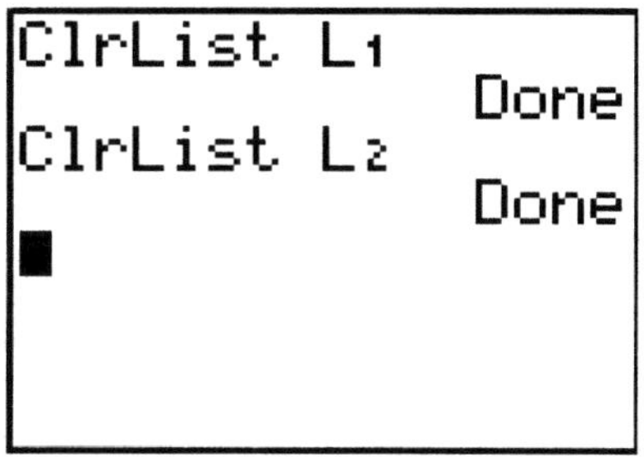

Clear List entries from the List screen:

- Go to the List screen.
- Press STAT EDIT and select 1: Edit.
- Overwrite the elements of the list or delete any elements you want using the [DEL] key.
- Note that the [CLEAR] key will not erase a highlighted value.

Clear an Entire List from the List screen:

- An entire list can be cleared in the STAT menu by pressing the up arrow until the top of the list is highlighted.
- Press [CLEAR] [ENTER] in this position to clear the contents of the entire list.
- If you press the [DEL] key when the column heading is highlighted, it will remove the list from the STAT editor, but the list will still possess its contents.

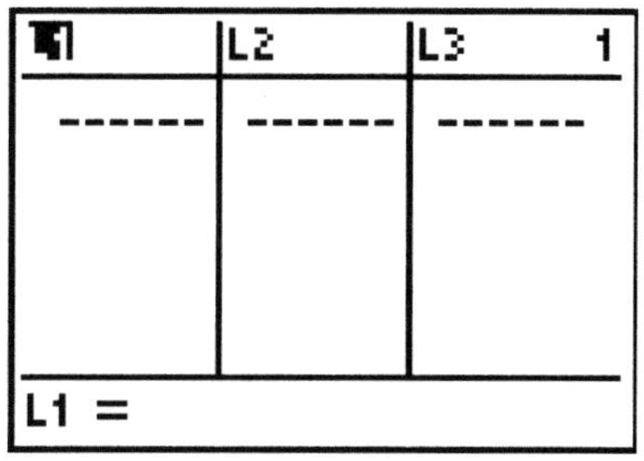

- The list name can be returned to the editor with [2nd] [DEL] (INS) command and typing in the list name.

Turn on a Plot

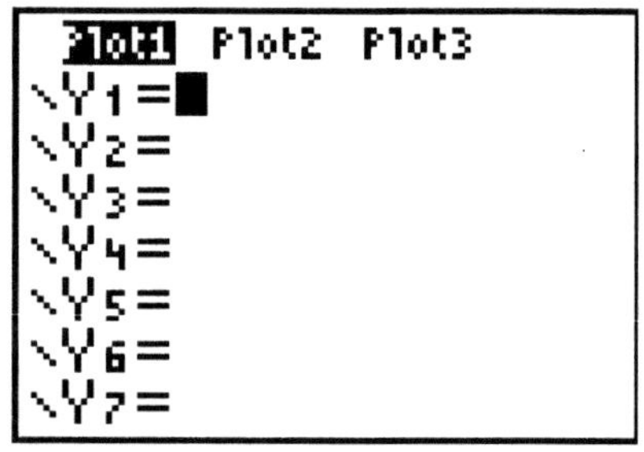

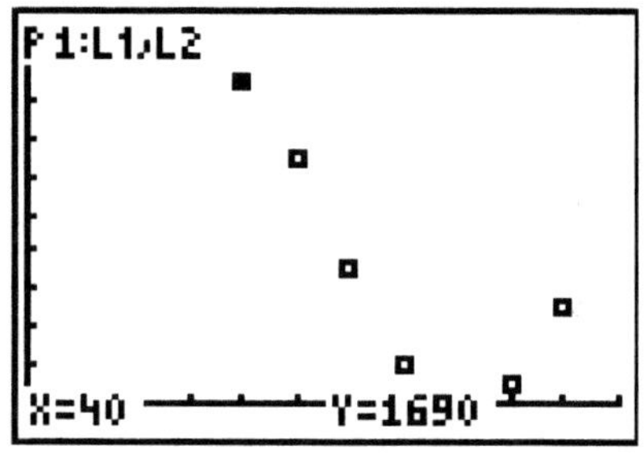

- Go back to the Y= menu and turn off or clear all the Y= functions, namely the ones that have a dark rectangle over the equal sign. To turn off a function, move the cursor over the *equal* sign of the function you wish to turn off.
- Press [ENTER] to remove the highlighted (dark) rectangle over the equal sign.
- Turn on Plot 1. On the graph screen, use the up arrow to go to `Plot 1`. Press [ENTER] to highlight it.
- Press [GRAPH] to see the points.
- Adjust the window if you wish to see the points more clearly.
- If you see something else, like a histogram, you must format the `Plot` using `STAT PLOT` as described below.

Use `STAT PLOT` to format the graph of data points

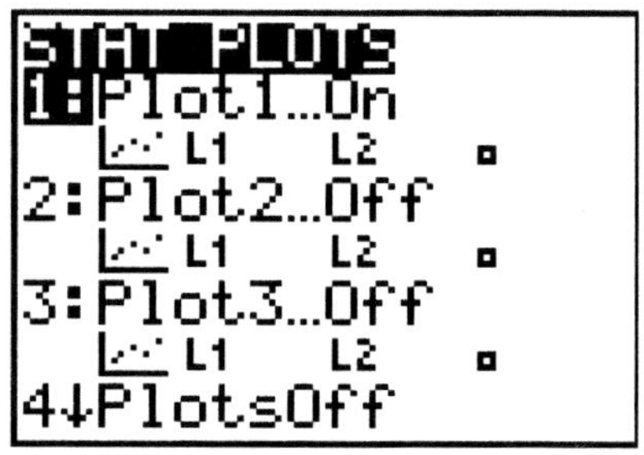

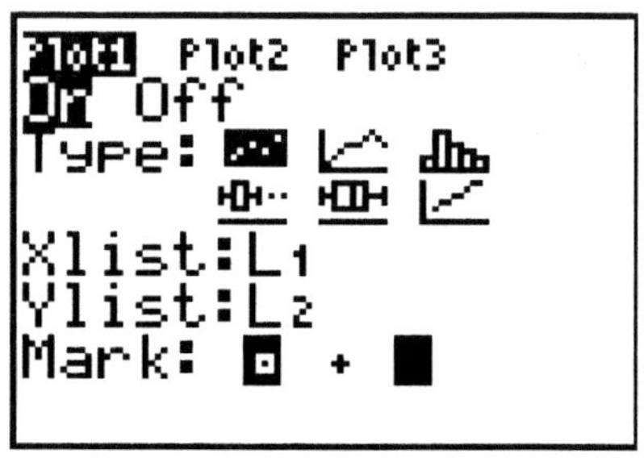

- Press [2nd] [Y=] to access STAT PLOT.
- Select 1: Plot 1...ON, press [ENTER].
- Select `Type`: first one for a scatter plot – individual points, press [ENTER].
- On the same screen select

  `XList:` L_1 or whichever list has the *x*-data

  `YList:` L_2 or whichever list has the *y*-data

  `Mark:` any one you wish

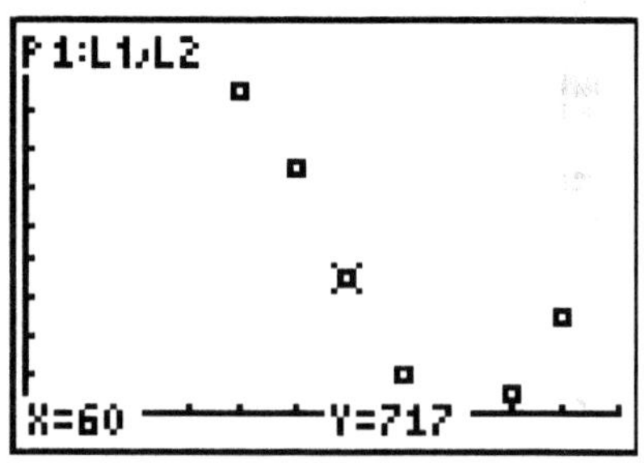

- Press the [GRAPH] key to see the scatter plot.
- Use [TRACE] and the right arrow to move from point to point. The *x* and *y* coordinates of each point are displayed at the bottom of the screen.
- `P1:L1,L2` is displayed in the upper left hand corner of the screen. This tells you that you are on `Plot 1`, the *x*'s are from list L_1, and the *y*'s are from list L_2.

1.2 Example 5 on the TI-86

Create a Scatter Plot of Data Points on the TI-86

Example 5 asks you to create a scatter plot of data points. Use the `STAT` menu of the calculator and define and select `Plot 1, Plot 2, or Plot 3.`

TABLE 1-12

Decade	Years from 1900 to End of Decade	Number of Executions
1930-1939	40	1690
1940-1949	50	1284
1950-1959	60	717
1960-1969	70	191
1970-1979	80	3
1980-1989	90	117
1990-1999	100	518

Set the Window

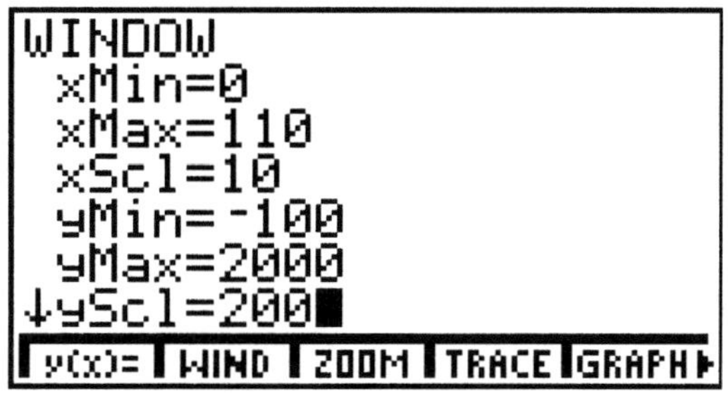

- First set the window.
- Press the [GRAPH] key, then [F2] to access WIND.
- Set the window elements to

  xMin=0, xMax=110, xScl=10, yMin= -100,

  yMax=2000, yScl=200.
- Press the down arrow or [ENTER] to move from line to line.
- Press [EXIT] to return to the Home screen.

Enter the Data Points

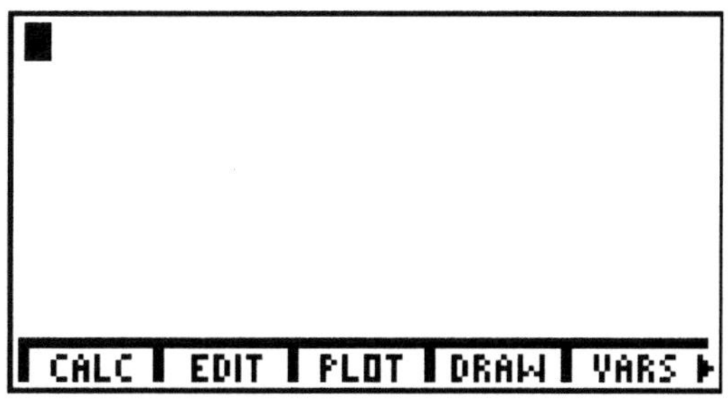

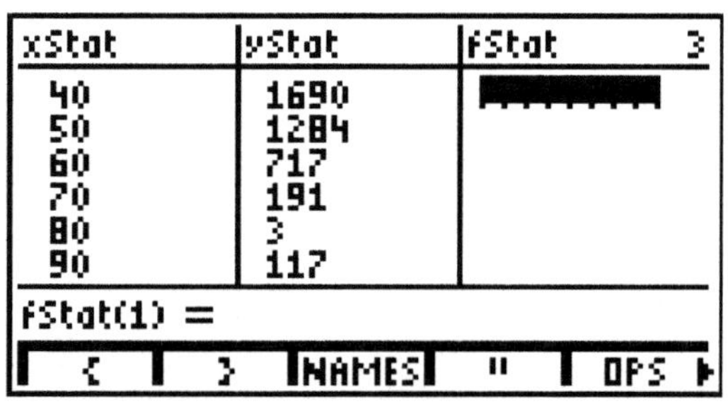

- Press the [2nd][+] keys to access the STAT menu.
- There are 6 menu items in the STAT menu. Press [MORE] to see the 6th item. Press [MORE] again to see the first 5 menu items.
- Press [F2] to select `Edit`. This brings you to the screen where you can enter the data in the lists.
- You may need to clear the lists. See directions below.
- Enter the *Years from 1900*-values in list xstat.
- Enter the *Number of executions*-values in ystat.
- Press [ENTER] after each entry.

Clearing Lists

- To clear entries from a list, overwrite the elements of the list or delete any elements you want using the [DEL] key.
- Note that the [CLEAR] key will not erase a highlighted value.
- An entire list can be cleared in the STAT menu by pressing the up arrow until the top of the list is highlighted. Press [CLEAR] [ENTER] in this position to clear the contents of the entire list.
- If you press the [DEL] key when the column heading is highlighted, it will remove the list from the STAT editor, but the list will still possess its contents in the NAMES menu. The list name can be returned to the editor by pressing [F3] to access the NAMES menu. Press the F key that is directly under the name you want to restore.

Turn on a Plot

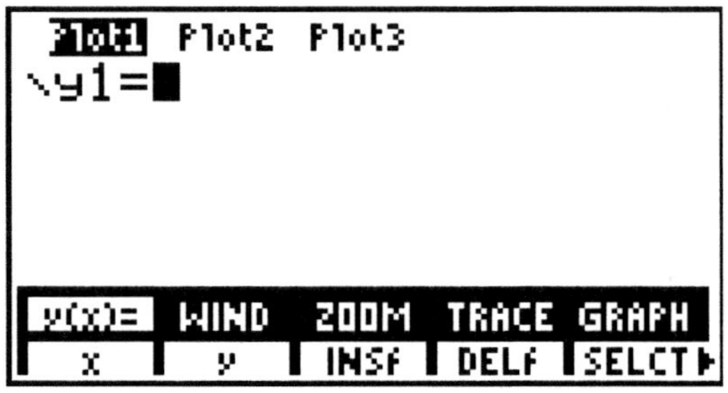

- Press the [GRAPH] key, then the [F1] key to go back to the y= menu. Turn off or clear all the y= functions, namely the ones that have a dark rectangle over the equal sign.
- To turn off a function, move the cursor to the function you wish to turn off. Press [F5] SELCT to change the function from on to off or vice versa.

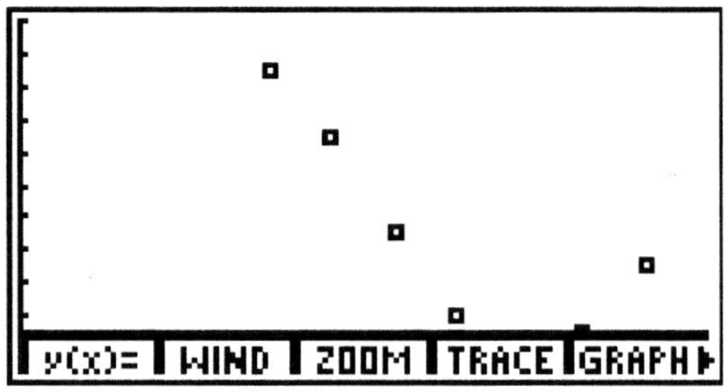

- Turn on Plot 1. On the graph screen, use the up arrow to go to Plot 1. Press [ENTER] to highlight it.
- If two menu lines are showing on the screen, press the [EXIT] key once to see only one line.
- Press [F5] GRAPH to see the points.
- Note that you may also graph the points by pressing [2nd] [F5] GRAPH when two menus lines are showing.
- Press [F4] TRACE to move from point to point.
- If you see something else, like a histogram, you must format the Plot using STAT PLOT as described on the following page.

Use STAT PLOT to format the graph of data points

- Press [2nd][+] keys to access the STAT menu.
- Press [F3] PLOT to format the plot.

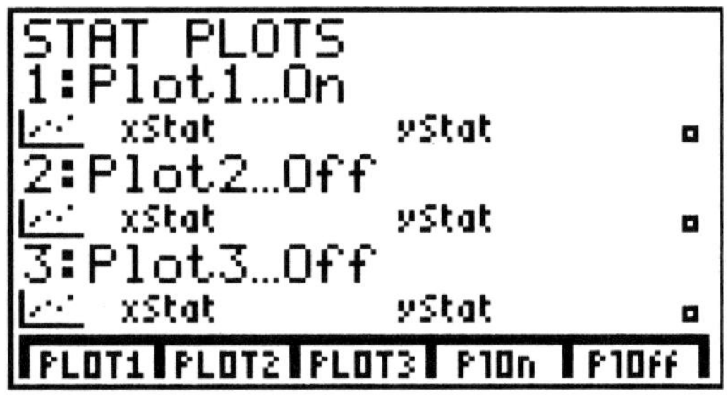

- Press [F1] to format Plot 1.
- Move the cursor to the word ON , then press [ENTER].
- Move the cursor to Type=.
- Press [F1] SCAT a scatter plot – individual points, press [ENTER].
- On the same screen select the F key below the name of the list that contains the *x*-data. We used xstat.
- On the same screen select the F key below the name of the list that contains the *y*-data. We used ystat.
- On the same screen select the F key below the Mark you wish to use for the points.
- Press [EXIT] [EXIT] to return to the Home Screen.

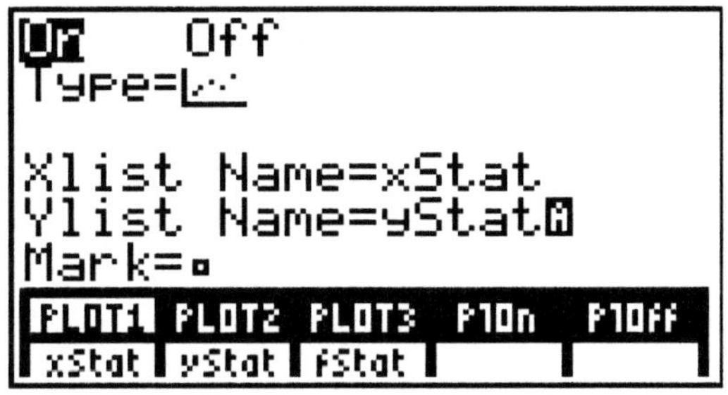

- Press the [GRAPH] key to see the scatter plot. You may also need to press [F5] GRAPH to see the scatter plot.
- Use [F4] TRACE and the right arrow to move from point to point. The *x* and *y* coordinates of each point are displayed at the bottom of the screen.

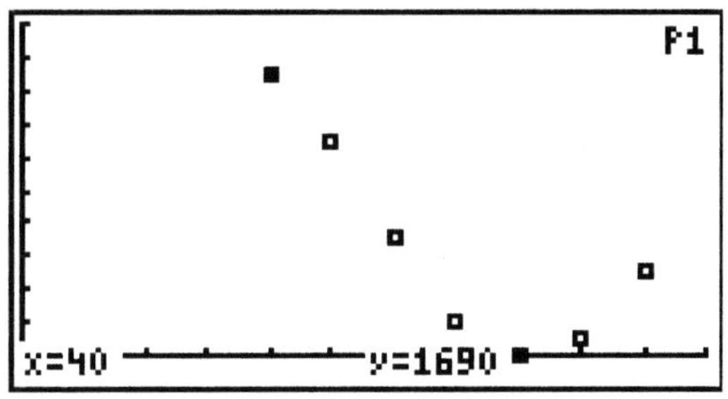

- P1 is displayed in the upper right hand corner of the screen. This tells you that you are on Plot 1.

1.2 Example 5 Using EXCEL

Create a Scatter Plot of Data Points in EXCEL

Example 5 asks you to create a scatter plot of data points.

To graph a function in EXCEL, you must first enter the data set, then use X-Y Scatter in the Chart Wizard to create the graph of the function.

Enter the Data Set

- Open a new worksheet in EXCEL.
- In cell A1, enter Years from 1900 to End of Decade, and in B1 enter the Number of Executions.
- To make the columns wider, go to the break between columns A and B. When the cursor looks like a cross, hold the left mouse button down and move the column to the right.
- Since the years are in increments of 10, we can create the data set quickly.
- In cell A2 enter the first value, 40.
- In cell A3 type the formula **=A2+10**. To do this by referencing the A2 cell, type the = sign in A3, then left click on the A2 cell.
- Click back into cell A3 and type +10.
- Use the lower right handle of the A3 cell to drag the formula through cell A8 where *years* = 100.
- Enter the values for the number of executions in each year in cells B2 thru B8 as shown.

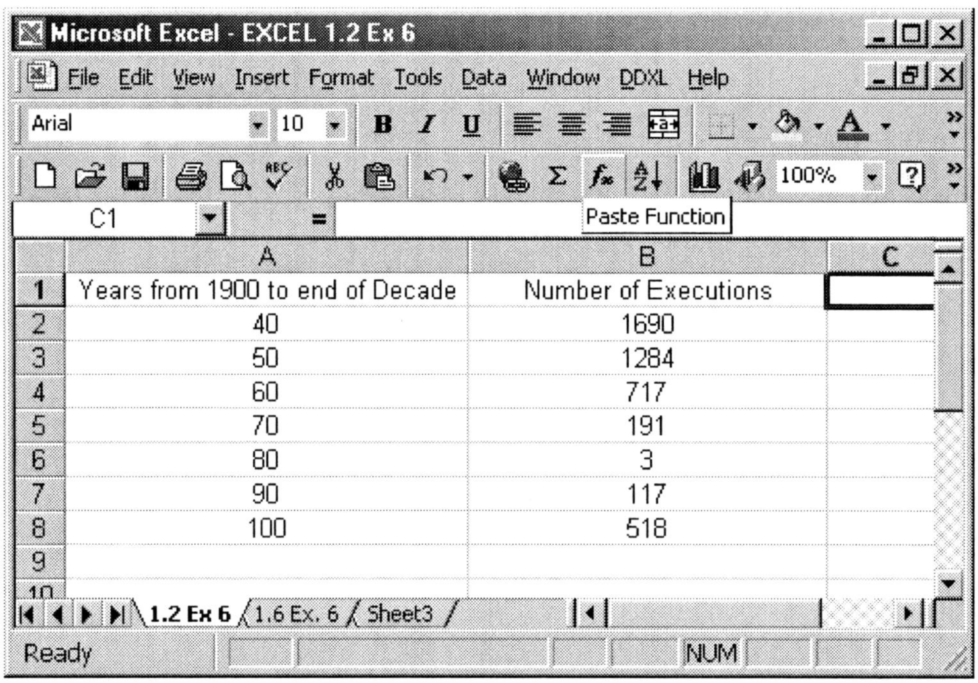

	A	B	C
1	Years from 1900 to end of Decade	Number of Executions	
2	40	1690	
3	50	1284	
4	60	717	
5	70	191	
6	80	3	
7	90	117	
8	100	518	
9			
10			

Graph the Scatter Plot

- To graph the function, first select the data set. To do this, left click on cell A1 and hold the left mouse button down until you highlight the entire table that contains the data.
- Select **Chart Wizard** on the menu bar.
- Select *XY* Scatter.
- Select **Scatter Compares pairs of values.** This is the first selection of scatter plot.

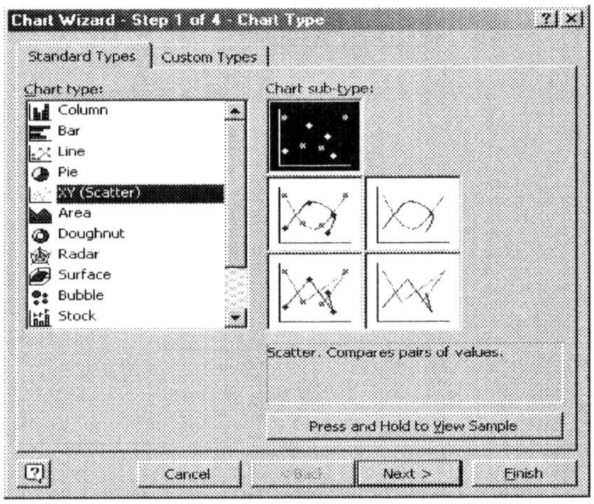

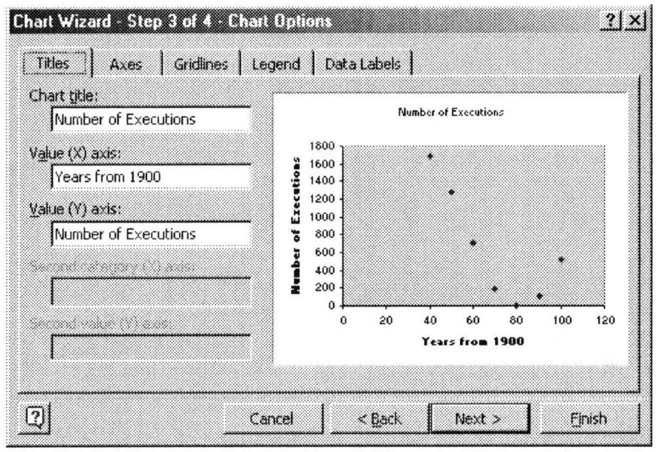

- Click on **NEXT.**
- Data range should be the range of cells you highlighted.
- Check Series in columns.
- Click on **NEXT.**
- Click on the **Titles t**ab and fill in the titles for the chart and the *x* and *y* axes. Enter the titles as shown.
- Click on the **Gridlines** tab and *uncheck* **major gridlines**.
- Click on **Legends** tab and *uncheck* **Show Legend.**
- Click on **NEXT.**
- Select As object in.

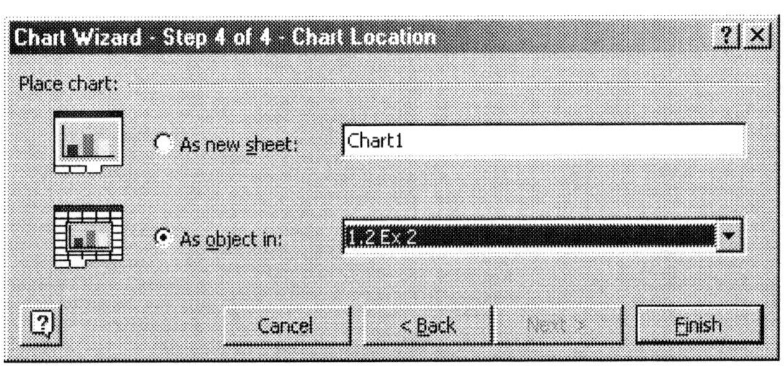

- Click on **FINISH.** The scatter plot will appear in EXCEL.
- You may right click on the chart to format the chart at any time.
- Note that when you click on the graph, the data set that was used to create the data is outlined.

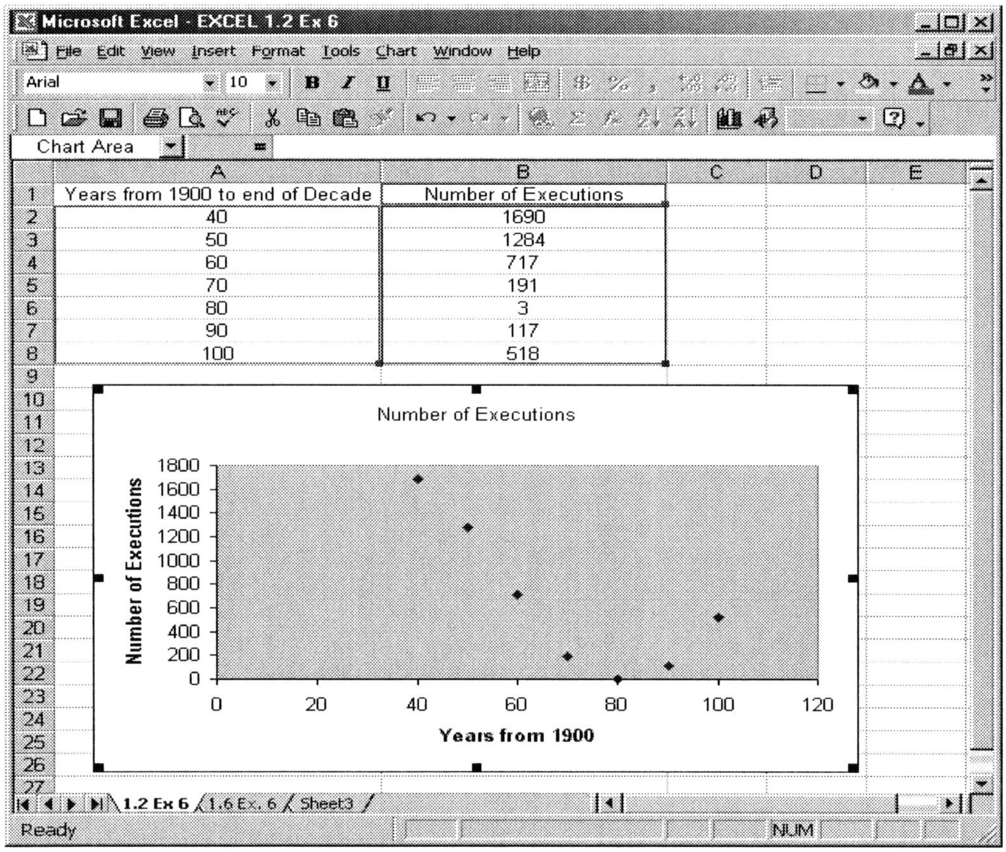

1.5 Example 5 on the TI-83, TI-83 PLUS, and TI-84 PLUS

Solve a Linear Equation Graphically Using the *x*-Intercept Method

Example 5 asks you to solve the equation $\frac{2x-3}{4}=\frac{x}{3}+1$.

First rewrite the equation with 0 on one side as $0=\frac{x}{3}+1-\frac{2x-3}{4}$, then enter the function as $\text{Y1}=x/3+1-(2x-3)/4$ on the calculator. Set the window as [-10, 15] by [-4,4] as shown below.

Enter the Equation $\frac{2x-3}{4}=\frac{x}{3}+1$

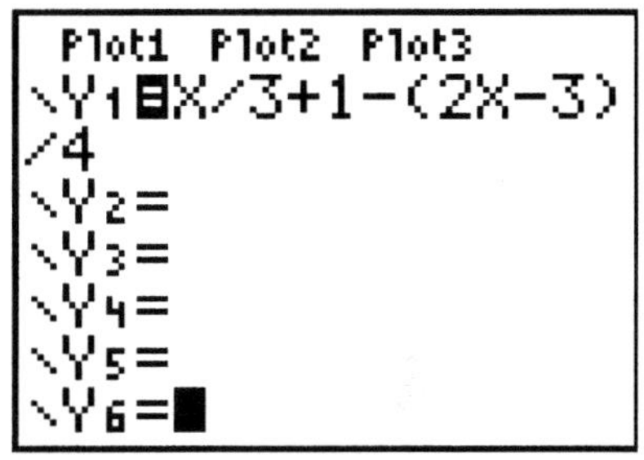

- Use the [Y=] key to enter the function as $\text{Y1}=x/3+1-(2x-3)/4$. Note that the parentheses in the second term are *essential.*
- Make sure that all other functions and plots are turned *off.*

Set the Window

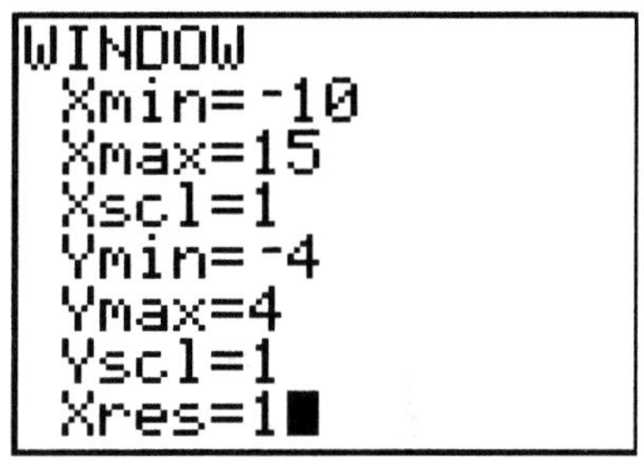

- Use the [WINDOW] key to set the window as shown.
- Set Xscl=1 and Yscl=1.

Graph the equation and use TRACE to estimate the zero

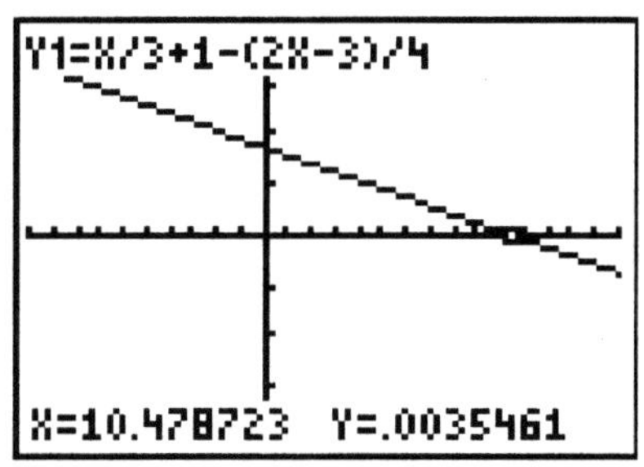

- Press the [GRAPH] key to see the graph.
- Press [TRACE] to see the equation.
- Use the right arrow [▸] key to get close to the *x*-intercept. Note that using [TRACE] does not necessarily give the *exact* value of the *x*-intercept.
- Here is x-intercept is close to x = 10.5.

Find the exact *x*-intercept

- Press [2nd] [TRACE] to access CALC. A CALCULATE screen appears.
- Press the [2] key or cursor down to 2:zero and press [ENTER].
- You are brought back to the graph screen.

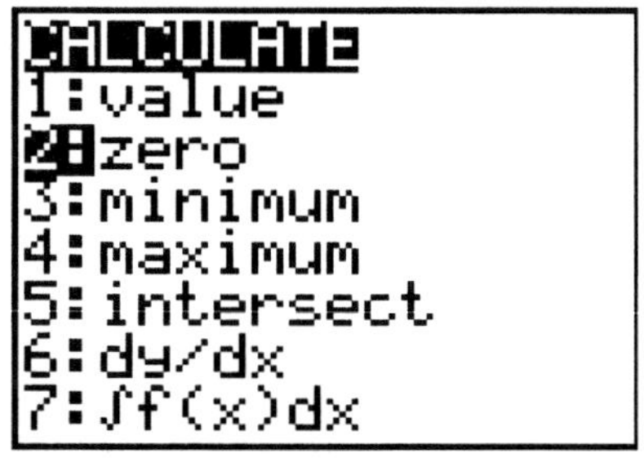

- You must select a *Left Bound* to tell the calculator where to look for the x-intercept. To do this, move the cursor somewhere close to the left side of the x-intercept. Press [ENTER].
- Notice that an arrow appears at the top of the screen above your left bound. This is the left bound of the interval that contains the x-intercept.

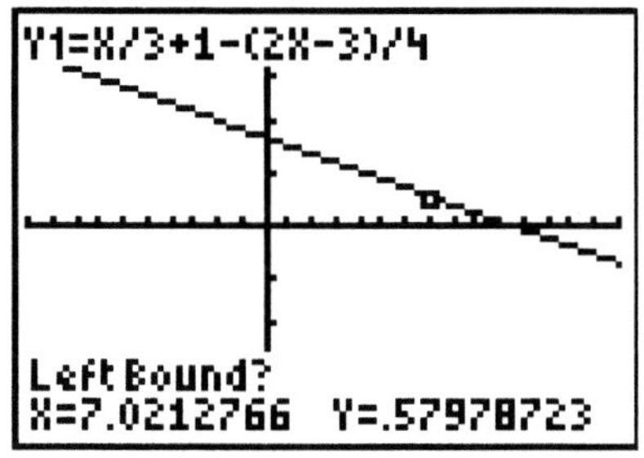

- You must select a *Right Bound* to tell the calculator where to look for the x-intercept. To do this, move the cursor somewhere close to the left side of the x-intercept. Press [ENTER].
- Notice that an arrow appears at the top of the screen above your right bound. This is the right bound of the interval that contains the x-intercept.

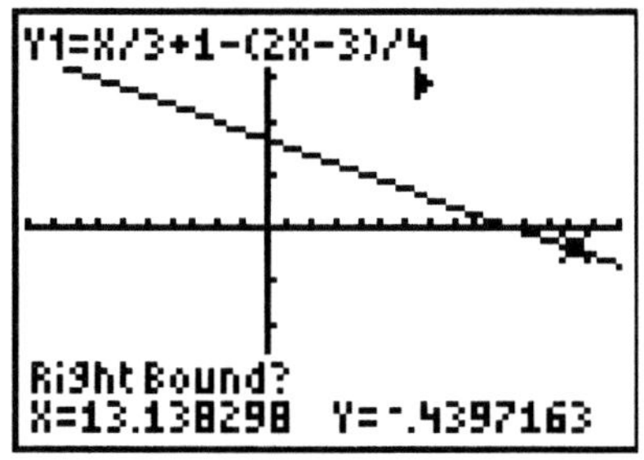

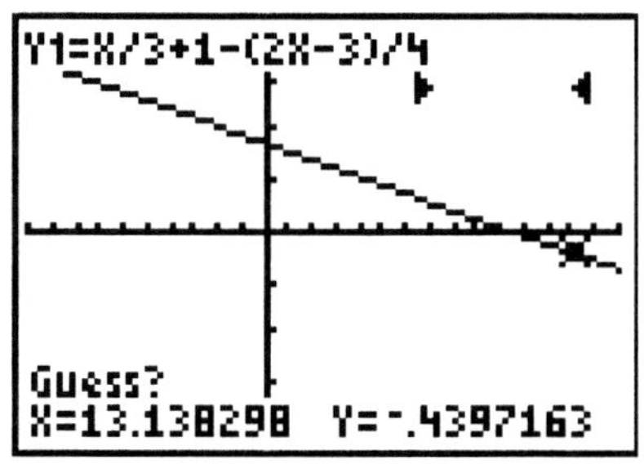

- Next, you must enter a *Guess*. You can enter a value somewhere within the interval defined by your left and right bounds. However, you can just press [ENTER].

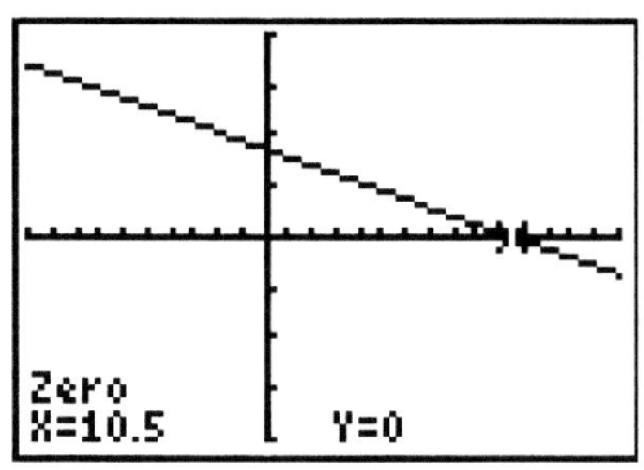

- Finally, press [ENTER] to see the coordinates of the x-intercept at the bottom of the screen.
- Note that the TI-83 and TI-83 PLUS call the x-intercept a *Zero*.

Solve a Linear Equation Graphically Using the Intersection Method on the TI-83, TI-83 Plus, and TI-84 PLUS

Example 5 asks you to solve the equation $\frac{2x-3}{4}=\frac{x}{3}+1$.

To use the Intersection Method, you must enter the left side of the function as Y1 $=(2x-3)/4$ and the right side of the function as Y2 $=x/3+1$ on the calculator. Set the window as [-4, 14] by [-2, 6] as shown below.

Enter the Equation $\frac{2x-3}{4}=\frac{x}{3}+1$

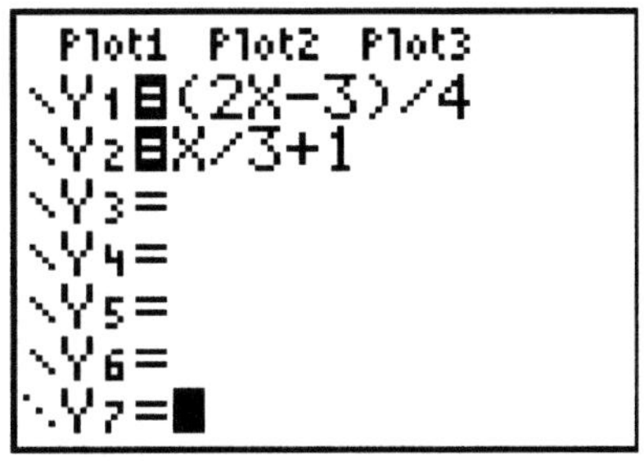

- Use the [Y=] key to enter the left side of the function as Y1 $=(2x-3)/4$ and the right side of the function as Y2 $=x/3+1$.
- Note that the parentheses in the numerator of Y1 are *essential.*

Set the Window

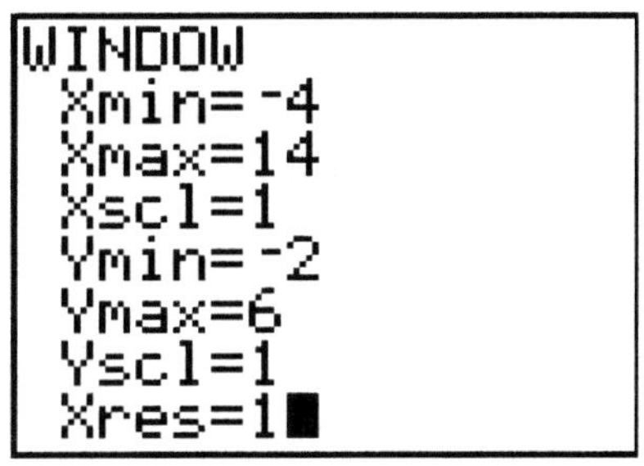

- Use the [WINDOW] key to set the window as shown.
- Set Xscl=1 and Yscl=1.

Graph the equations

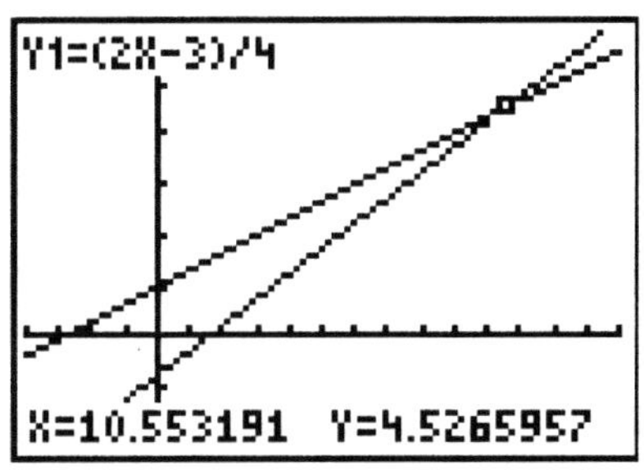

- Press the [GRAPH] key to see the graphs.
- Press [TRACE] to see the equation of one of the lines. Use the up or down arrow to see the equation of the second line.
- Use the right arrow [▶] key to get close to the *intersection point* of the two lines. Note that using [TRACE] does not necessarily give the *exact* coordinates of the *intersection point*.

Find the exact coordinates of the Intersection point

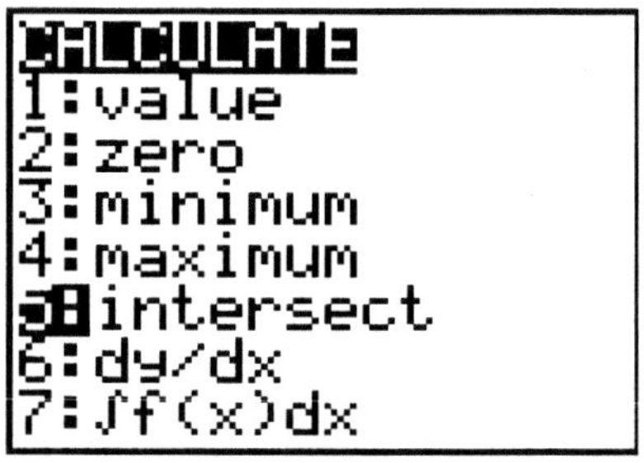

- Press [2nd] [TRACE] to access CALC. A CALCULATE screen appears.
- Press the [5] key or cursor down to 5:intersect and press [ENTER].
- You are brought back to the graph screen.

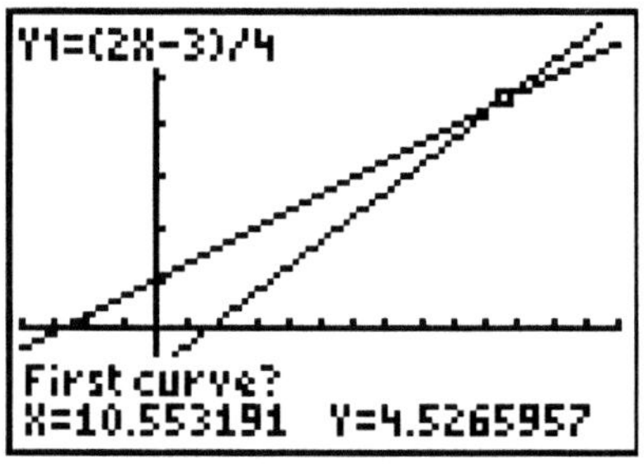

- You must select a *First curve* to tell the calculator which of the lines you want to use for the intersection. Press [ENTER].

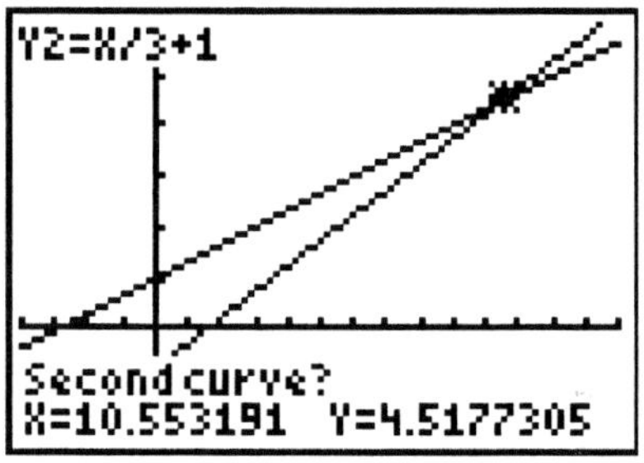

- Next you must select a *Second curve* to tell the calculator which second curve to use for the intersection. Use the up or down arrow key to move to the other line. Press [ENTER].

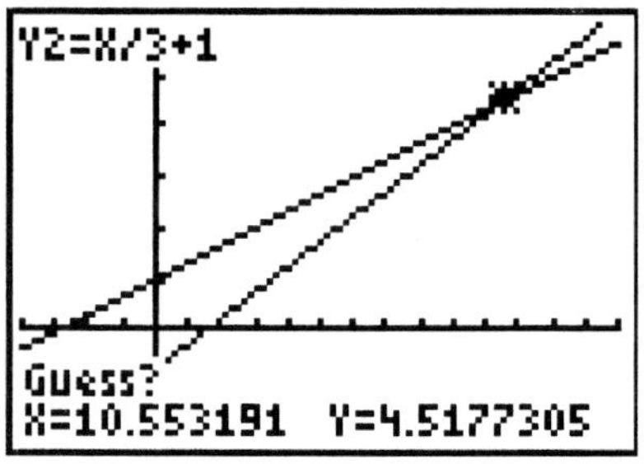

- Next, you must enter a *Guess.* You can enter a value or you can just press [ENTER].

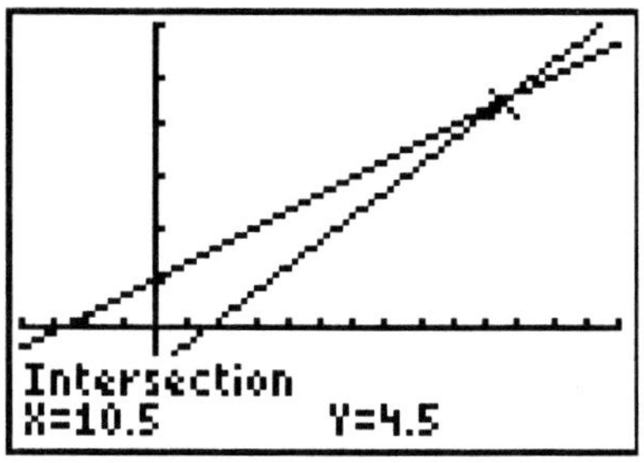

- Finally, press [ENTER] to see the coordinates of the *Intersection* at the bottom of the screen.

1.5 Example 5 on the TI-86

Solve a Linear Equation Graphically Using the *x*-Intercept Method on the TI-86

Example 5 asks you to solve the equation $\dfrac{2x-3}{4}=\dfrac{x}{3}+1$.

First rewrite the equation with 0 on one side as $0=\dfrac{x}{3}+1-\dfrac{2x-3}{4}$, then enter the function as $\text{y1}=x/3+1-(2x-3)/4$ on the calculator. Set the window as [-10, 15] by [-4,4] as shown below.

Enter the Equation $\dfrac{2x-3}{4}=\dfrac{x}{3}+1$

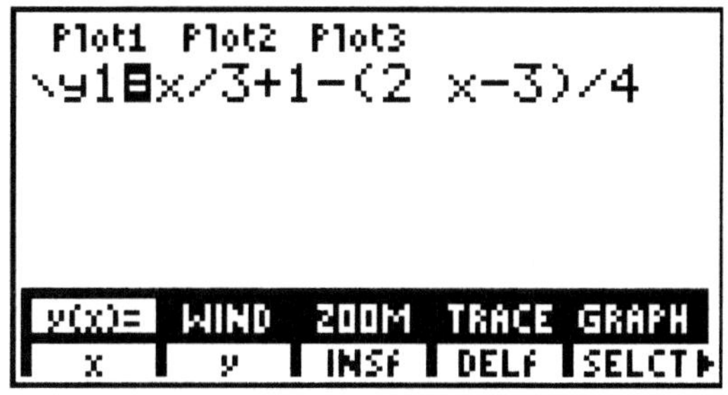

- Press the [GRAPH] key, then press [F1] to access the *y(x)=* menu. Enter the function as $\text{y1}=x/3+1-(2x-3)/4$. Note that the parentheses in the second term are *essential.*
- Make sure that all other functions and plots are turned *off.*

Set the Window

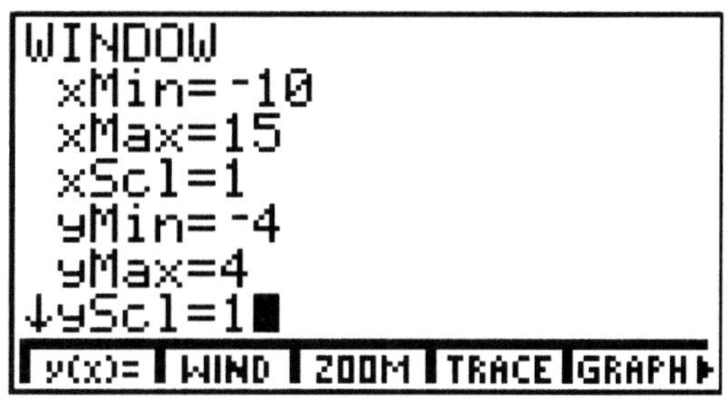

- Press [EXIT] then the [F2] key to set the window as shown.
- Set xScl=1 and yScl=1.

Graph the equation

- Press the [F5] key to select GRAPH to see the graph.
- Press [F4] to TRACE along the curve.
- Use the right arrow [▶] key to get close to the *x*-intercept. Note that using TRACE does not necessarily give the *exact* value of the *x*-intercept.
- Note that the 1 in the upper right corner tells you that you are looking at the graph of y1.

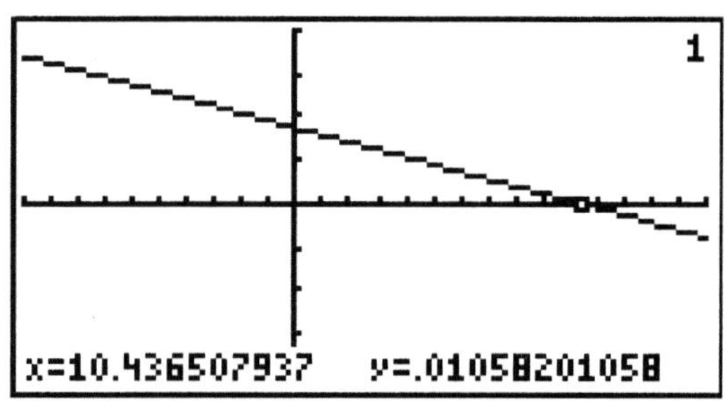

Find the exact *x*-intercept

- Press [EXIT] to see the Graph menu again.
- Press [MORE] [F1]to access the MATH menu of the graph screen.
- Press the [F1] key to select ROOT.

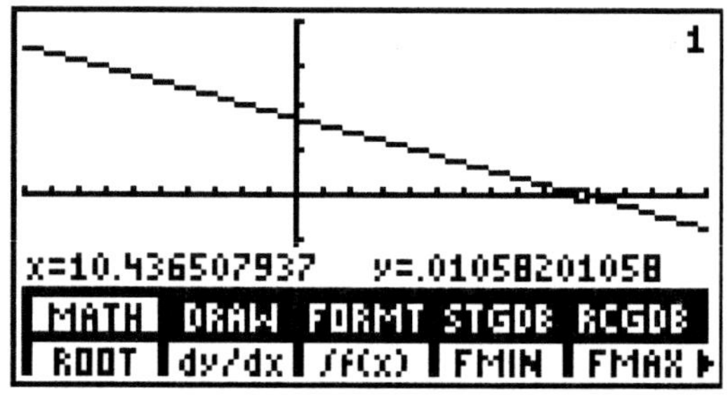

- You must select a *Left Bound* to tell the calculator where to look for the *x*-intercept. To do this, move the cursor somewhere close to the left side of the *x*-intercept.
- Press [ENTER].

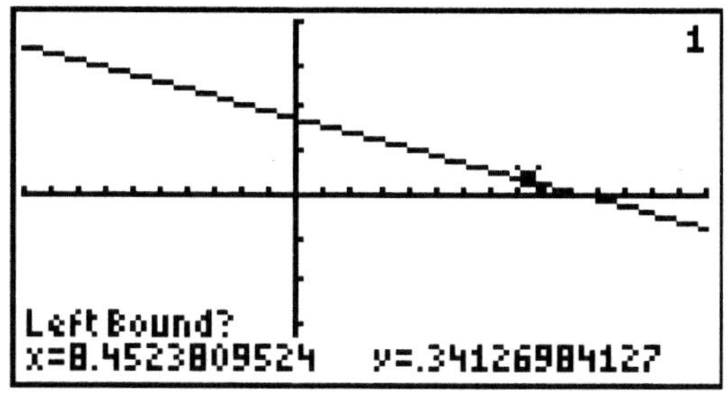

- Notice that an arrow appears at the top of the screen above your left bound. This is the left bound of the interval that contains the *x*-intercept.
- You must select a *Right Bound* to tell the calculator where to look for the *x*-intercept. To do this, move the cursor somewhere close to the right side of the *x*-intercept.
- Press [ENTER].

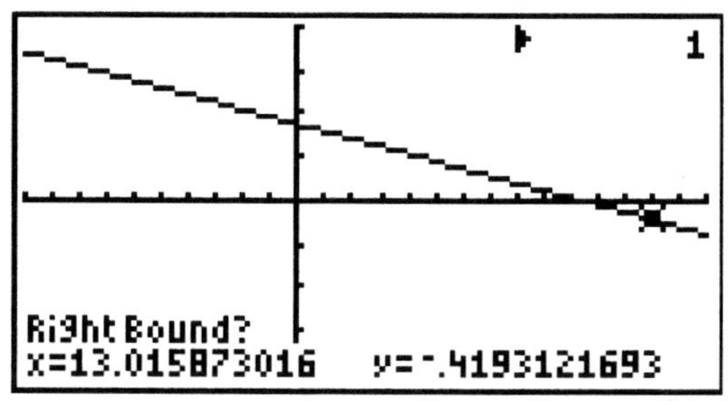

- Notice that an arrow appears at the top of the screen above your right bound. This is the right bound of the interval that contains the *x*-intercept.
- Next, you must enter a *Guess*. You can enter a value somewhere within the interval defined by your left and right bounds. However, you can just press [ENTER].

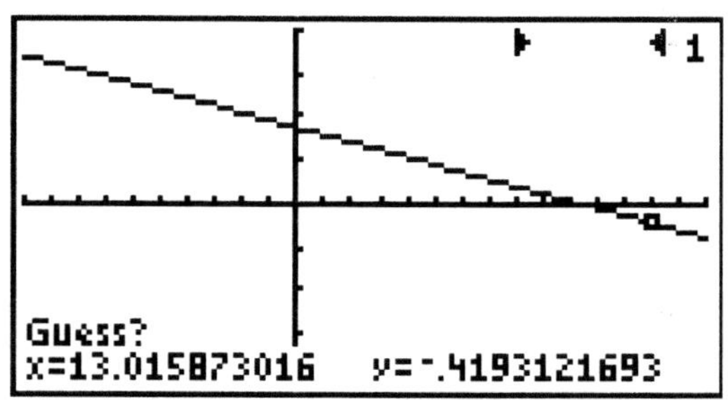

- The coordinates of the *x*-intercept are shown at the bottom of the screen.
- Note that the TI-86 calls the *x*-intercept a *Root*.

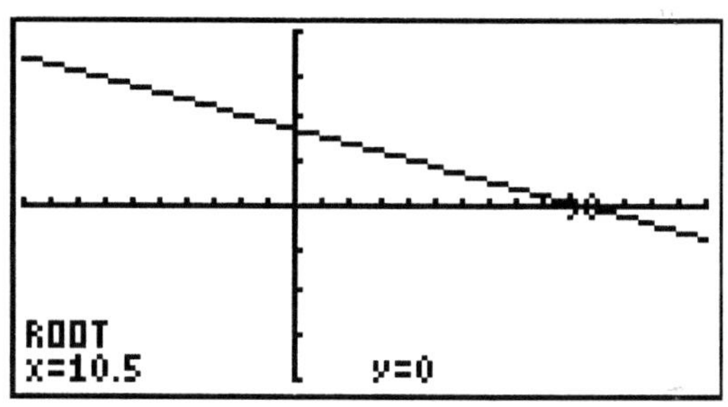

Solve a Linear Equation Graphically Using the Intersection Method on the TI-86

Example 4 asks you to solve the equation $\frac{2x-3}{4}=\frac{x}{3}+1$.

To use the Intersection Method, you must enter the left side of the function as y1 $=(2x-3)/4$ and the right side of the function as y2 $=x/3+1$ on the calculator. Set the window as [-4, 14] by [-2, 6] as shown below.

Enter the Equation $\frac{2x-3}{4}=\frac{x}{3}+1$

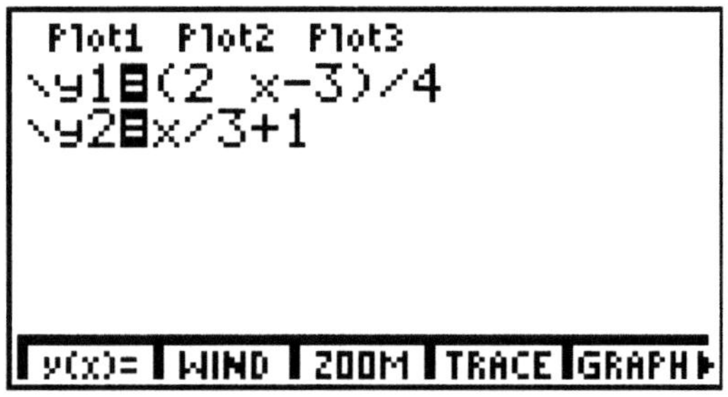

- Press the [GRAPH] key, then press [F1] to access the y(x)= menu. Enter the left side of the equation as y1 $=(2x-3)/4$ and the right side of the function as y2 $=x/3+1$.
- Note that the parentheses in the numerator of y1 are *essential.*
- Press [EXIT] to display one menu line rather than two menu lines.

Set the Window

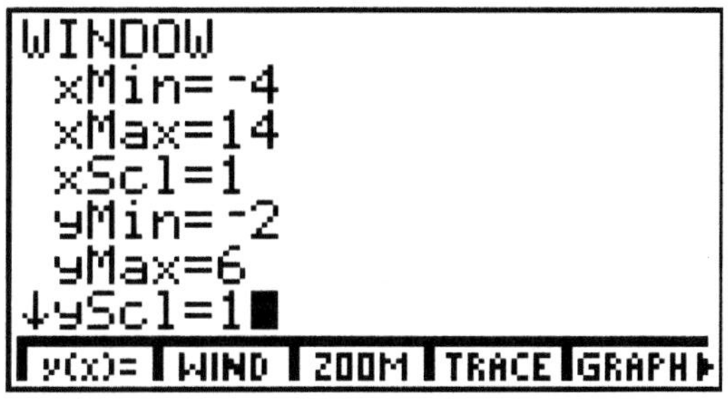

- Press the [F2] WIND key to set the window as shown.
- Set xScl=1 and yScl=1.

Graph the equations

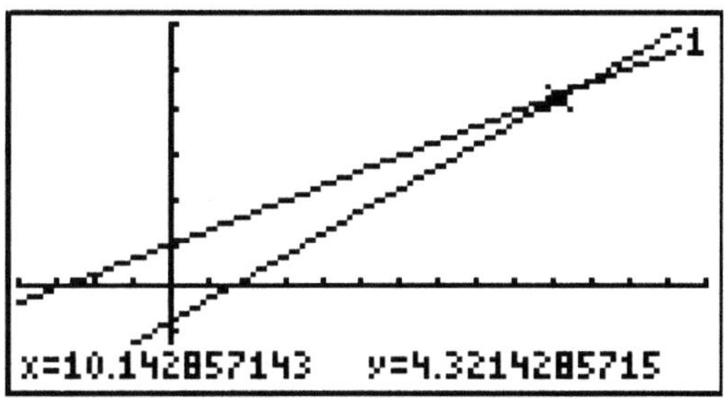

- Press the [F5] GRAPH key to see the graph.
- Press [F4] TRACE to walk along the curves.
- Use the up or down arrow to move to the second line. The number in the upper right hand corner tells you which equation you are on.
- Use the right arrow [▸] key to get close to the *intersection point* of the two lines. Note that using TRACE does not necessarily give the *exact* coordinates of the *intersection point.*

Find the exact coordinates of the Intersection point

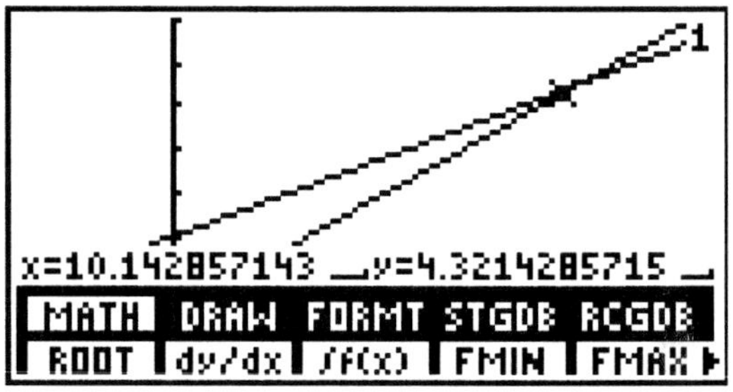

- Press [EXIT] to see the Graph menu again.
- Press [MORE] [F1] to access the MATH menu of the graph screen.

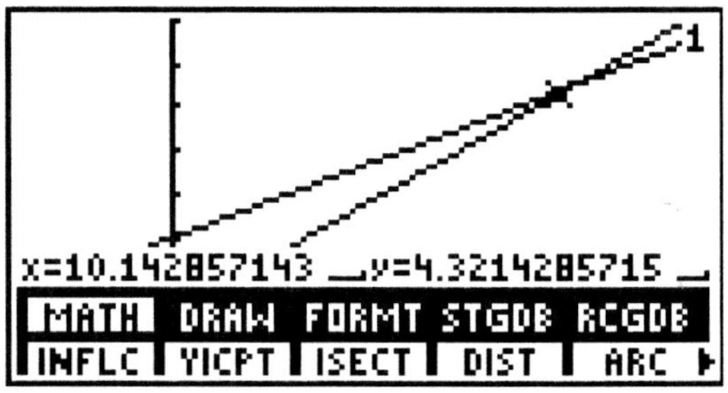

- Press the [MORE] key again.
- Press the [F3] key to select ISECT.
- This will locate the intersection point of the two lines.

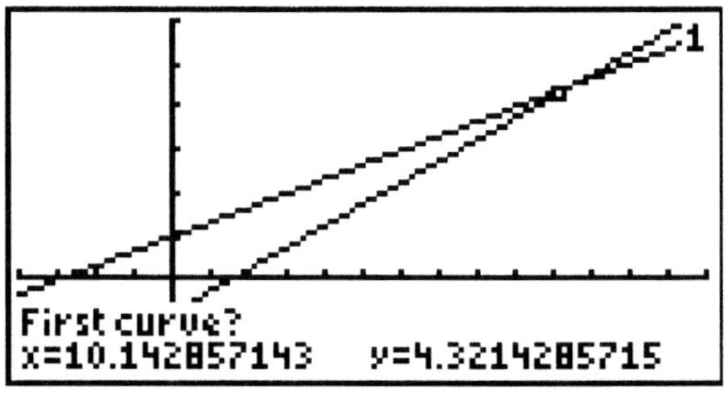

- You must select the *First curve* to tell the calculator which line you want to use first. The number in the upper right corner of the screen tells you which curve you are on. The 1 says you are on y1 and the 2 says you are on y2.
- Press [ENTER] to select the first curve.

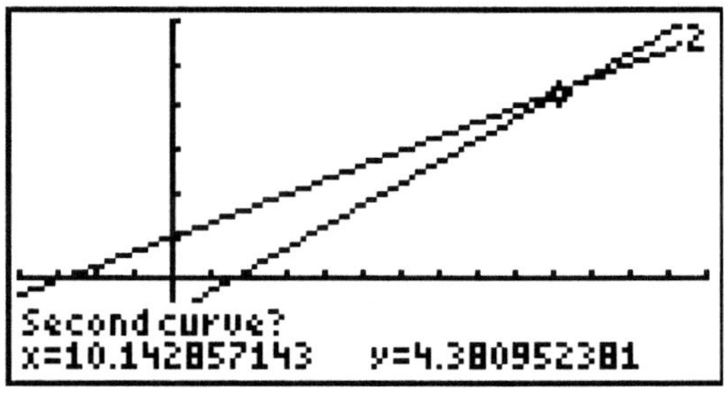

- Next you must select a *Second curve* to tell the calculator which second curve to use for the intersection. Use the up or down arrow to move to the second line. Notice that the cursor is now on the other line.
- Press [ENTER].

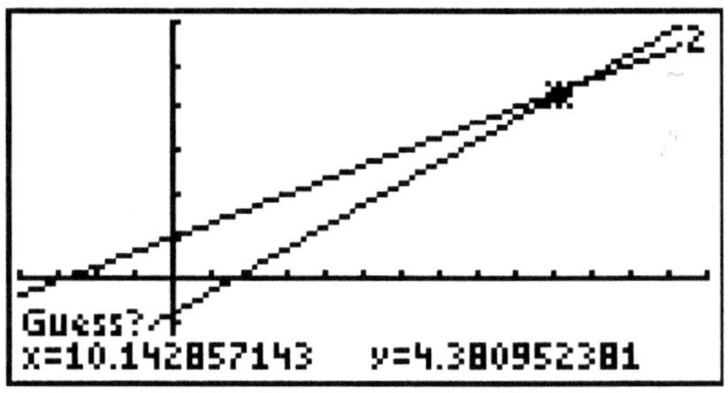

- Next, you must enter a *Guess*. You can enter a value or you can just press [ENTER].

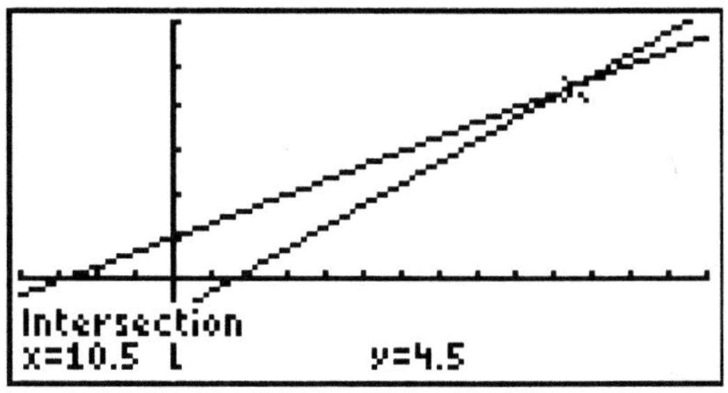

- Finally, you see the coordinates of the *intersection point* at the bottom of the screen.

1.5 Example 6 on the TI-83, TI-83 PLUS, and TI-84 PLUS

Follow the directions from Example 5 to solve using the x-intercept method and to solve using the intersection method.

1.5 Example 6 on the TI-86

Follow the directions from Example 5 to solve using the x-intercept method and to solve using the intersection method.

1.5 Example 6 using EXCEL

Find the *x*-intercept of a graph using GOAL SEEK in EXCEL.

Example 6 asks you to find the x-intercept of the function $120 = 0.55x - 2.886$. First subtract 120 from both sides of the equation to get the equation $0 = 0.55x - 122.886$. This is the equation you will use for Goal Seek.

To use Goal Seek in EXCEL you must first enter any value of x and the equation.

- Open a new worksheet.
- First enter a label for the first two columns, say x in cell A1 and y in cell B1.
- Enter any value of x, say 1, into cell A2 and the equation as $= 0.55 * A2 - 122.886$ into cell B2.

  Note that you can click on cell A2 to put it in the formula.
- Press **ENTER.** You will see the value –122.336 in cell B2.

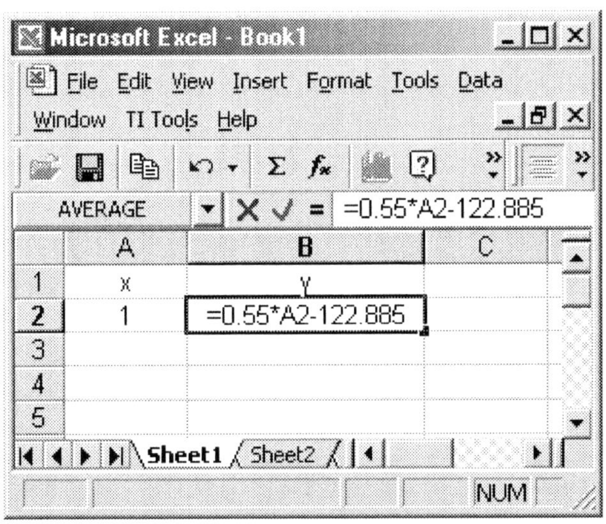

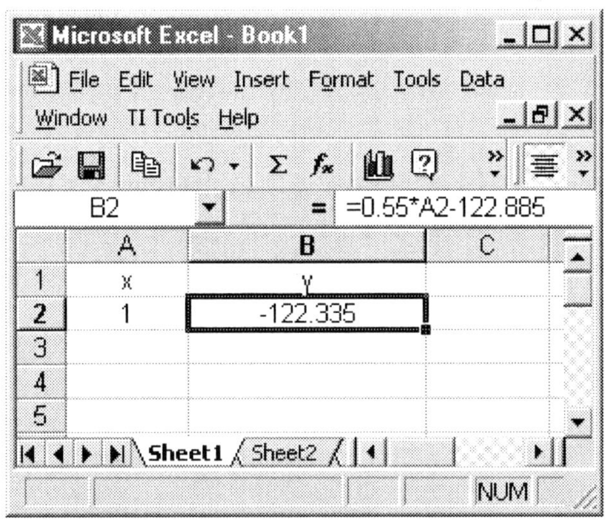

- Place the cursor back into B2, the cell that contains the formula.
- Go to **Tools** on the toolbar.
- Select **Goal Seek.** A dialog box appears.

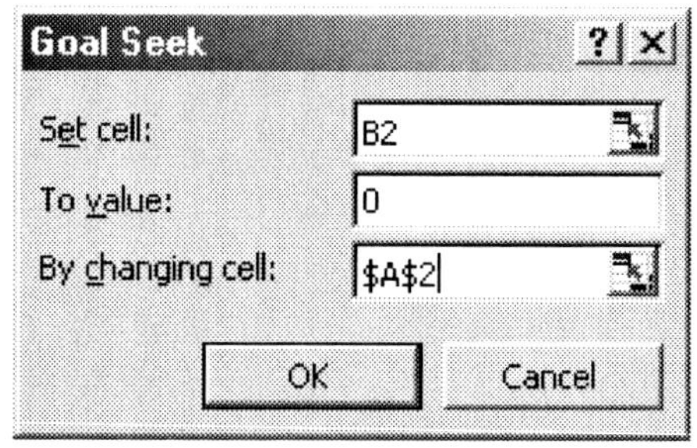

- Fill in the dialog box as follows:

  Set cell: B2 Note that you can just click in cell B2.

  To value: 0

  By changing cell: A2 Note that you can just click in cell A2 to put it here.

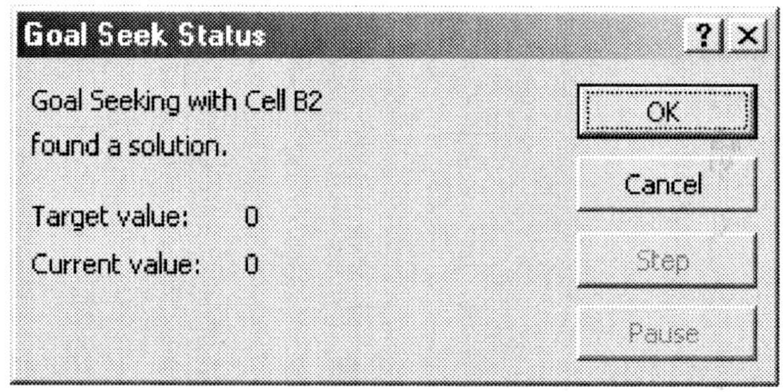

- Click **OK.**
- Another dialog box appears that tells you a solution has been found.
- Click **OK.**

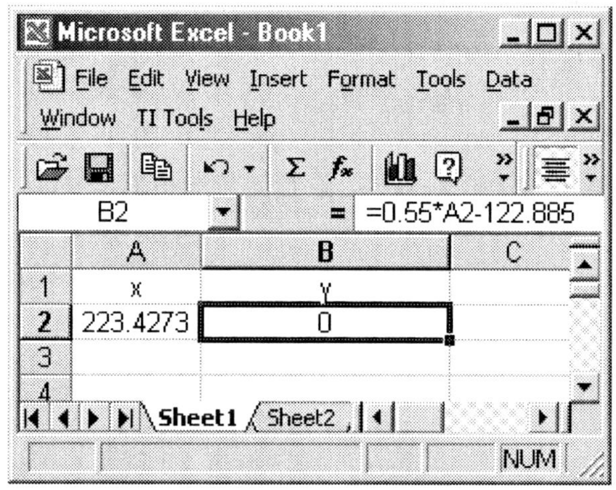

- Cell A2 now contains the value 223.4273 and cell B2 contains 0.
- Your solution is that the x-intercept is 223.4273. Notice that the equation you used is in the menu bar as shown below.

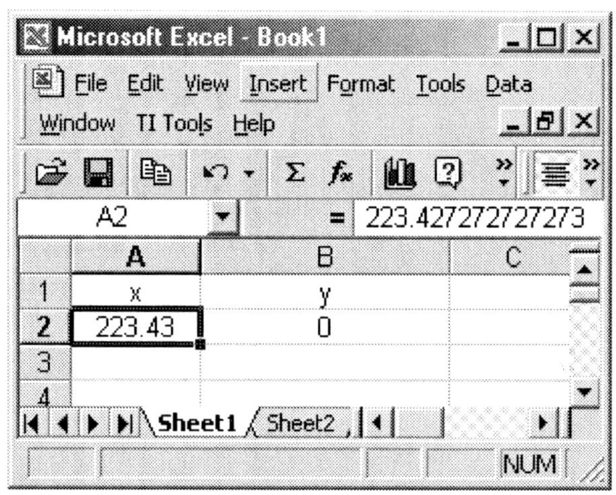

- Click in cell A2 to round to 2 decimal places.
- Go to **Format** in the menu bar, select **Cells.**
- Select **Number** and select 2 decimal places as shown. Click **OK.**

Find the exact coordinates of the Intersection point using GOAL SEEK in EXCEL.

Example 6 asks you to find the solution of the function $120 = 0.55x - 2.886$ by finding the intersection of two lines, $y1 = 0.55x - 2.886$ and $y2 = 120$. First find the difference of the two equations, $y2 - y1$ to get the equation $0 = 0.55x - 122.886$. This is the equation you will use in Goal Seek.

Note that this equation, $0 = 0.55x - 122.886$, is exactly the same function that you needed to use with x-intercept method described on the previous pages. Follow the same directions for using **Goal Seek.**

- First enter a label for the first two columns, say *x* in cell A1 and *y* in cell B1.
- Enter any value of *x*, say 1, into cell A2 and the equation as $= 0.55 * A2 - 122.886$ into cell B2. Note that you can click on cell A2 to put it in the formula.
- Press **ENTER.** You will see the value –122.336 in cell B2.
- Go to **Tools** on the toolbar.
- Select **Goal Seek.** A dialog box appears.
- Fill in the dialog box as follows:

  Set cell: B2 Note that you can just click in cell B2.

  To value: 0

  By changing cell: A2 Note that you can just click in cell A2 to put it here.
- Click **OK.** Another dialog box appears that tells you a solution has been found. Click **OK.**

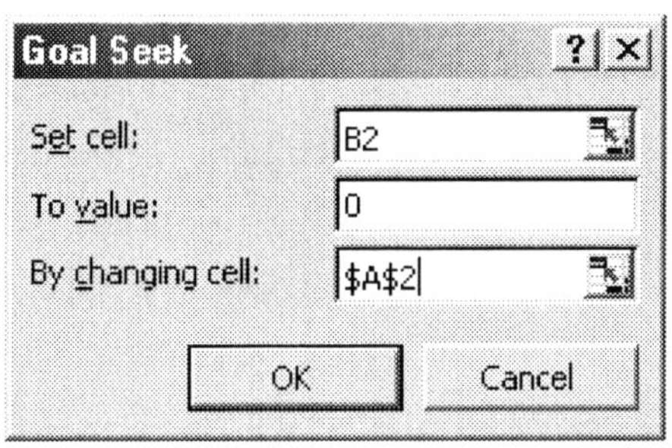

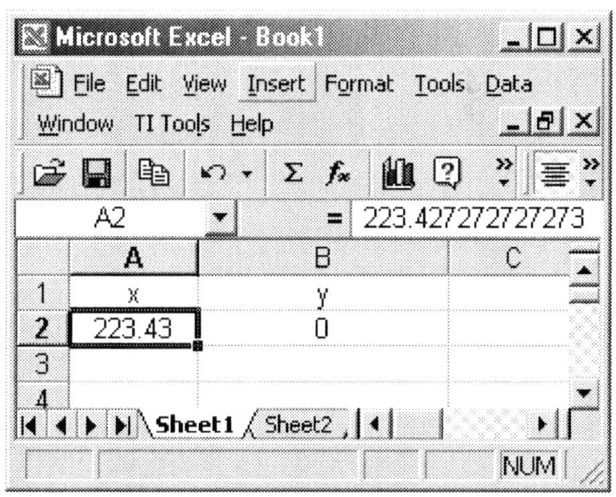

- Cell A2 now contains the value 223.43 and cell B2 contains 0.
- Your solution is that the *x*-coordinate of the intersection point is 223.43.

1.6 Example 1 on the TI-83, TI-83 PLUS, and TI-84 PLUS

Example 1 of section 1.6 *Health Service Employment* gives a table of the numbers of full and part-time employees in offices and clinics of dentists in the selected years between 1970 and 1998. You are asked to create a scatter plot, graph some potential lines that may fit the data, and then find the linear regression line. Finally, you must graph the linear regression line and the scatter plot on the same grid.

Year	1970	1975	1980	1985	1990	1995	1998
x = # years past 1970	0	5	10	15	20	25	28
# Employees (in thousands)	222	331	415	480	580	644	666

Create the scatter plot - See 1.2 Example 5 for detailed instructions.

- Set the window to [-5, 30] by [100, 800] as shown or use [ZOOM] 9: ZoomStat to set the window for the data.
- Use [STAT] EDIT 1: Edit to enter the *x* values into L1 and the # employees into L2.
- Clear the lists if needed. Cursor to the name of the list to be cleared, press [CLEAR] [ENTER].
- Press [2nd] [Y=] to turn on and format Plot 1. Be sure that the data is coming from L1 and L2.
- Press the [GRAPH] key and then [TRACE] to see the scatter plot. Use the right and left arrows to move from point to point.
- Use [Y=] to enter the three linear equations given as shown. You may look at them one at a time by turning off the other two, or you can look at all 3 equations at the same time to decide which equation is a better fit for the data points. If you look at all 3 equations at once, use the up and down arrow to move from line to line. Keep Plot 1 on.

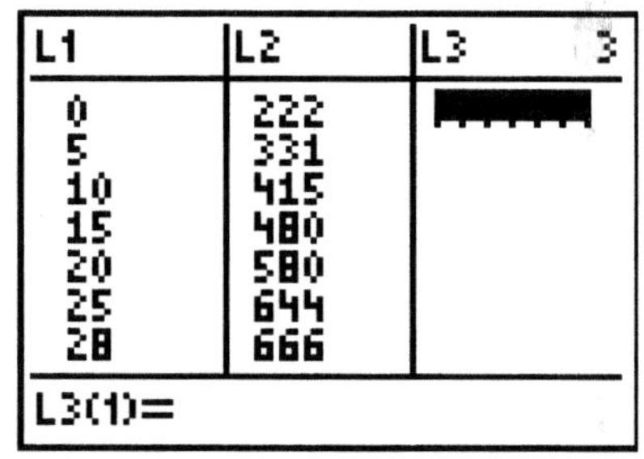

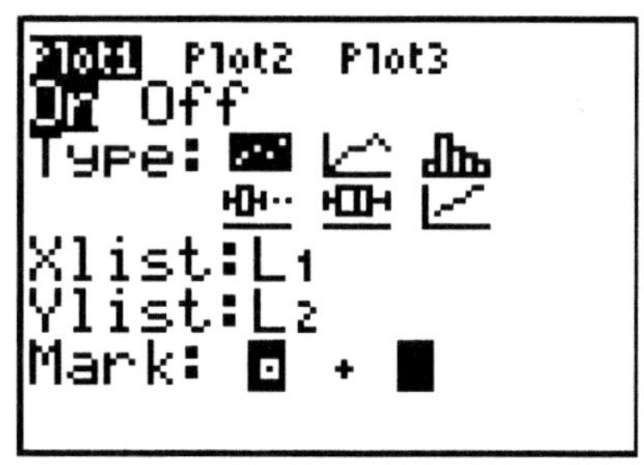

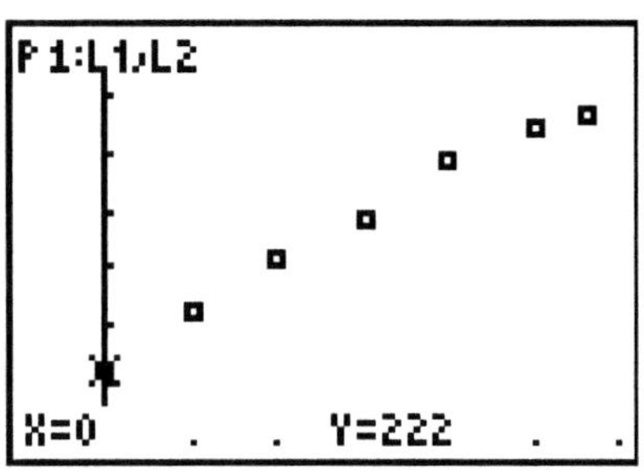

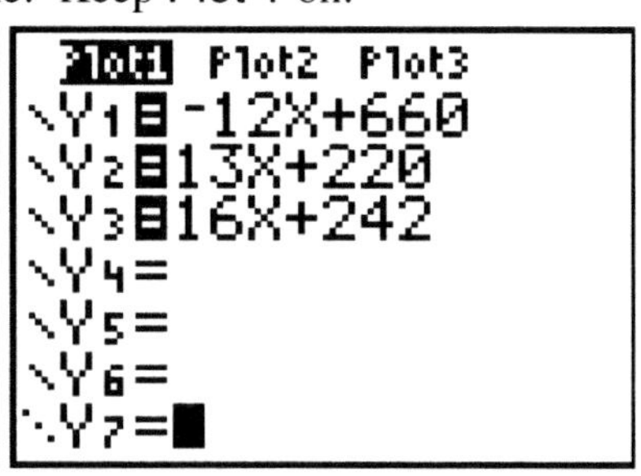

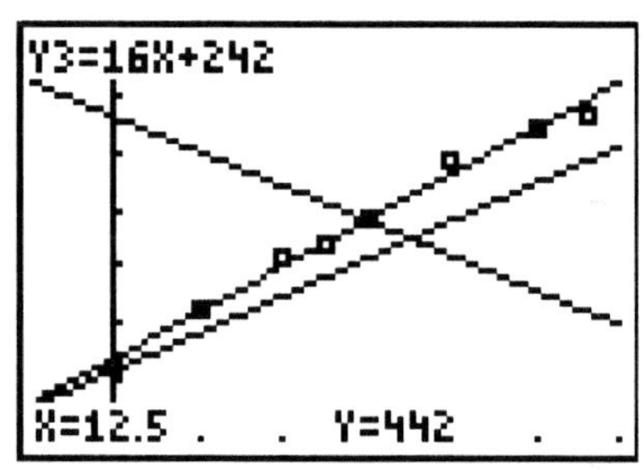

- Press [GRAPH] then [TRACE] and the up arrow [▲] until you see the Y3 $=16x+242$ equation. This line is the best fit of the 3 equations given.

Find the Linear Regression Equation that is the "Best Fit" for the Data.

The linear regression or "best fit" equation for the data in this example is given as $y = 15.910x + 242.758$. Note that this equation is rounded to 3 decimal places. Here is how to find the linear regression equation on the TI-83, TI-83 PLUS and TI-84 PLUS.

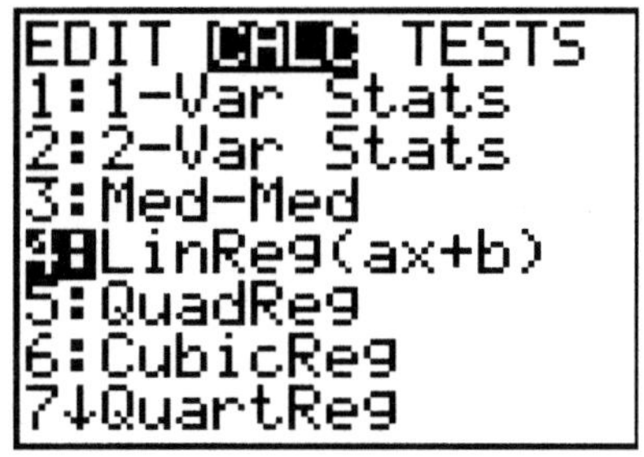

- The data is already entered into L1 and L2, and Plot 1 is formatted for scatter plot.
- Turn off or clear Y1, Y2, and Y3.
- Next, press the [STAT] key. Use the right arrow to go to CALC.
- Press [4] to select 4:LinReg(ax+b). This puts LinReg(ax+b) on the Home screen.
- Press [ENTER] to see the values of a and b as shown. Hence, the equation of the line through the data points is $y = 15.90963588x + 242.7582149$.

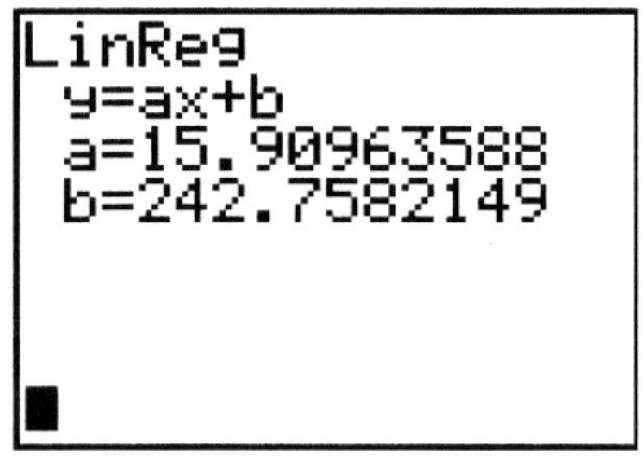

- Use the [MODE] key and Float 3 to round these values to 3 decimals as shown.

- Press [CLEAR] to return to the Home screen. Press [ENTER] to see the Linear Regression equation in 3 decimal places, $y = 15.910x + 242.758$.

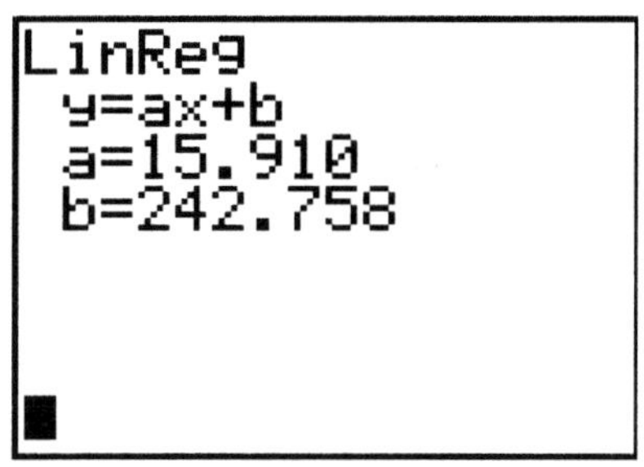

Display r and r^2

You may not see the values for the linear correlation coefficient *r*. To do so, you must go to the Catalog and set DiagnosticOn.

- Press [2nd] [0] to access the catalog. This is a complete listing of all the functions available on the calculator. You are now in alpha mode (see the A in the upper right hand corner of the screen.
- Press the [x^{-1}] to jump to the D's.
- Use the down arrow to find DiagnosticOn. Be sure that the arrow is next to it.

- Press [ENTER] twice until you see the word Done on the Home Screen.

- Redo finding the linear equation. The screen will now also include r and r^2.

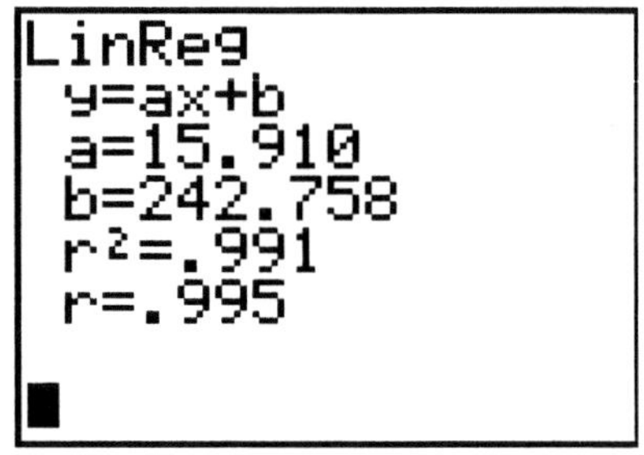

Graph the Linear Regression Equation on the Graph Screen

We would like to graph the linear regression equation on the same screen as the scatter plot.

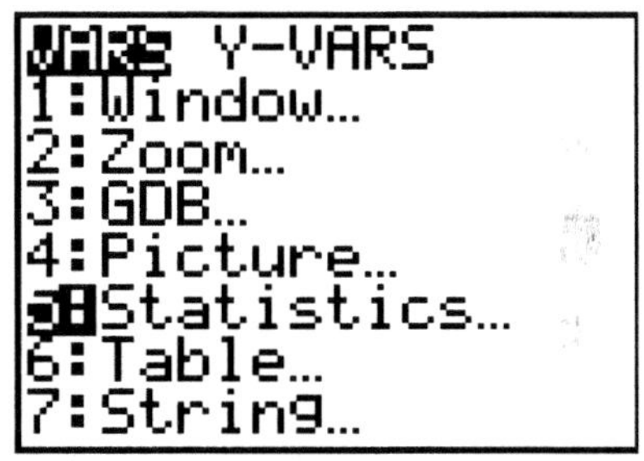

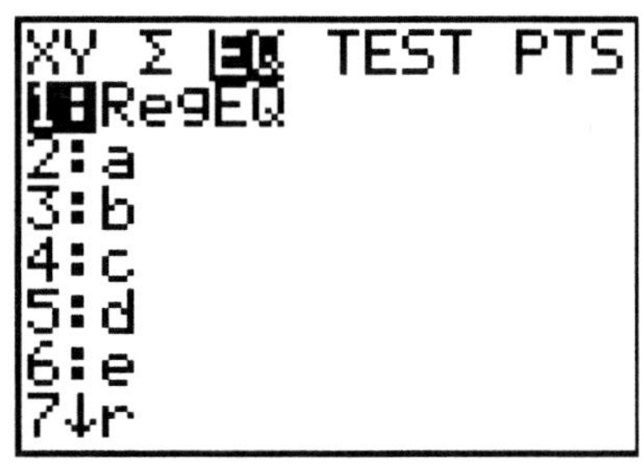

- Turn off or clear all Y= functions. Be sure that Plot 1 is on.
- Move the cursor to the Y= line you wish to use.
- Press the [VARS] key.
- Select 5:Statistics.
- Move the right arrow twice to see the EQ menu.
- Select 1:RegEQ.
- Press [ENTER]. This puts the regression equation into the Y= menu.
- Press the [GRAPH] key to see the regression equation and the scatter plot on the same screen.
- Press [TRACE] to trace the graphs.
- Use the up or down arrow to toggle between the scatter plot and the linear regression equation.

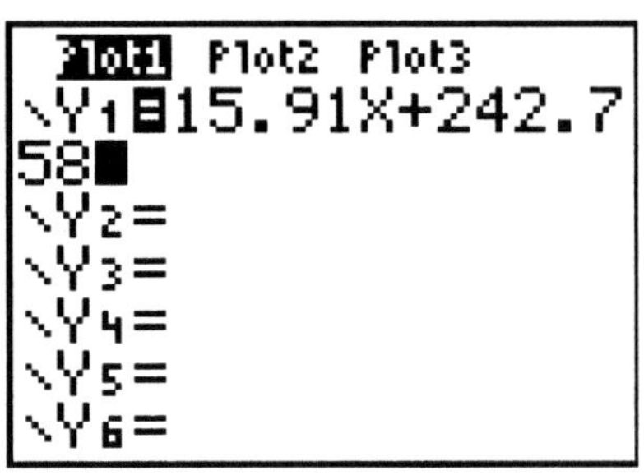

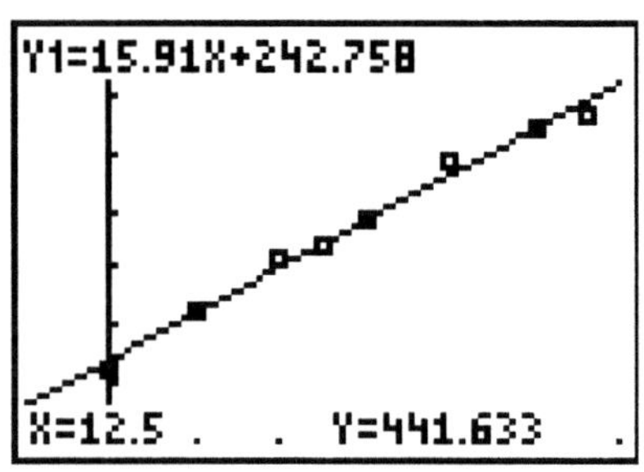

1.6 Example 1 on the TI-86

Example 1 of section 1.6 *Health Service Employment* gives a table of the numbers of full and part-time employees in offices and clinics of dentists in the selected years between 1970 and 1998. You are asked to create a scatter plot, graph some potential lines that may fit the data, and then find the linear regression line. Finally, you must graph the linear regression line and the scatter plot on the same grid.

Year	1970	1975	1980	1985	1990	1995	1998
x = # years past 1970	0	5	10	15	20	25	28
# Employees (in thousands)	222	331	415	480	580	644	666

Set up the STAT editor to include L1, L2, …, L6.

Before we do this example, let's set up STAT EDIT to look more like the lists in the TI-83 calculators. We must insert a new column one at a time. The new column is inserted to the left of an existing column, so we will create the new columns in reverse order L6, L5, L4, L3, L2, and L1.

- Press [2nd] [+] to access the STAT menu.
- Press [F2] to select EDIT.
- Use the up arrow to put the cursor on the first list xstat.
- Press [2nd][DEL] to insert a new column.

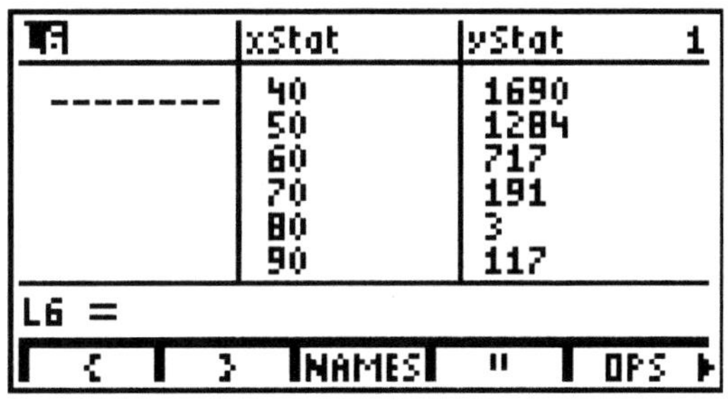

- At the bottom of the screen you are asked to give the new column a name.
- You are in alpha mode, so press [7] [ALPHA] [6] to input L6.
- Press [ENTER].
- Insert columns L5, L4, L3, L2, L1 in the same way.
- All 6 new columns are now included in NAMES. Use the [MORE] key to see all the columns that are now available to you.

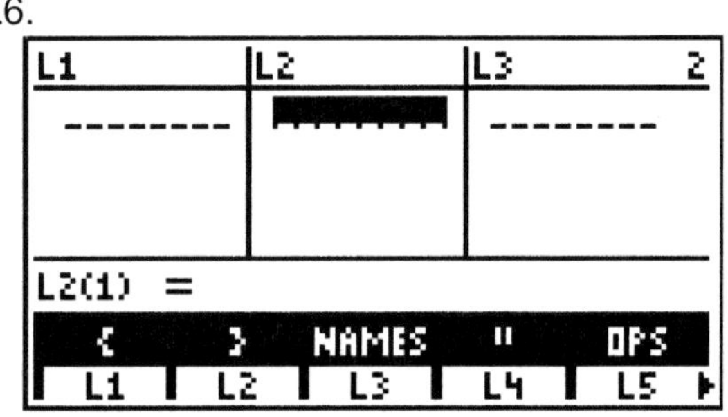

Create the scatter plot - See 1.2 Example 5 for detailed instructions.

- Set the window to [-5, 30] by [100, 800] or use [F3] ZOOM then [MORE] [F5] ZDATA to set the window for the data.
- Use [2nd] [+] [F2] to see the STAT EDIT screen. Enter the *x* values into L1 and the # employees into L2.

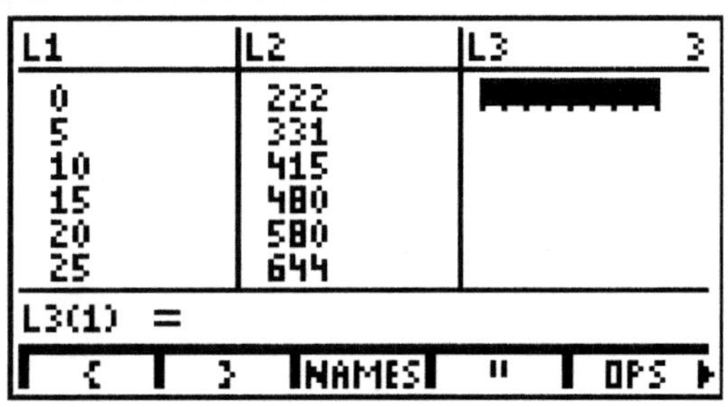

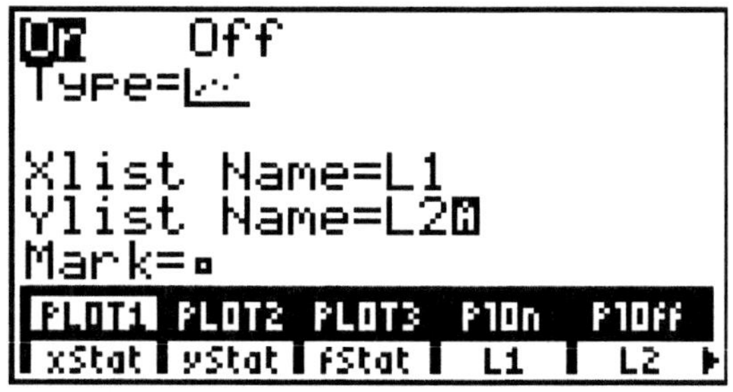

- Clear the lists if needed. Cursor to the name of the list to be cleared, press [CLEAR] [ENTER].
- Press [EXIT] [2nd] [+] to see the STAT menu.
- Press [F3] PLOT then [F1] to format PLOT1 as shown. Be sure that the data is coming from L1 and L2.

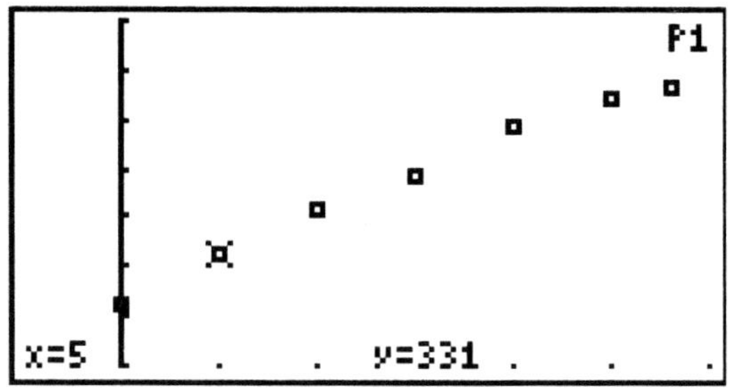

- Press [EXIT] [EXIT] to return to the Home screen.
- Press the [GRAPH] key and then [F5] to see the scatter plot.
- Press [TRACE] to see the coordinates of the points. Use the right and left arrows to move from point to point.

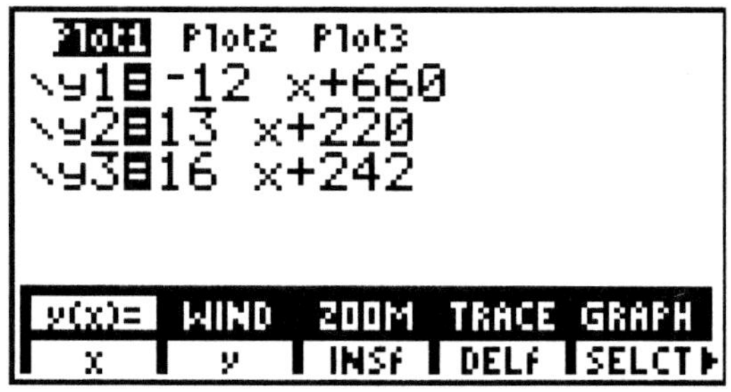

- Press [EXIT][F1]to enter the three linear equations given as shown. You may look at them one at a time by turning off the other two, or you can look at all 3 equations at the same time to decide which equation is a better fit for the data points. If you look at all 3 equations at once, use the up and down arrow to move from line to line. Keep Plot 1 on.

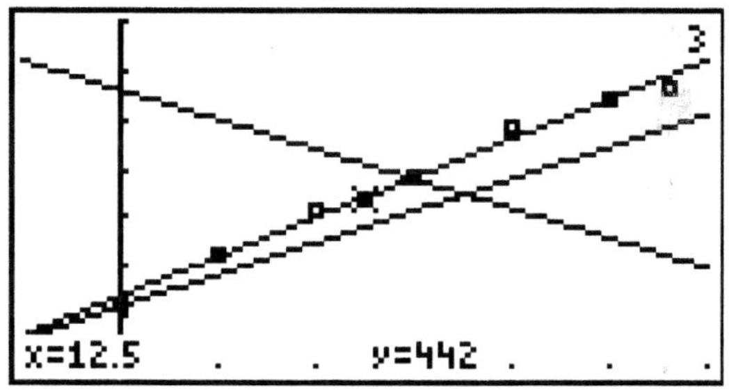

- Press [EXIT] [F5] to see the graphs. Then use [F4] TRACE and the up arrow [▲] until you see the y3 $= 16x + 242$ equation. You are on this equation when you see the numeral 3 in the upper right hand corner of the screen. This is the line that is the best fit of the 3 equations given.

Find the Linear Regression Equation that is the "Best Fit" for the Data.

The linear regression or "best fit" equation for the data in this example is given as $y = 15.910x + 242.758$. Note that this equation is rounded to 3 decimal places. Here is how to find the linear regression equation on the TI-86.

- The data is already entered into L1 and L2, and Plot 1 is formatted for scatter plot.
- Clear y1, y2, and y3. Place the cursor on y1.
- Press [EXIT] until you are on the Home Screen.

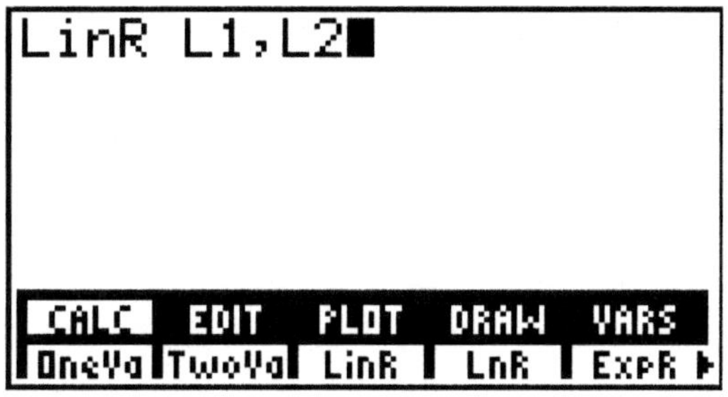

- Go back to the STAT menu. Press [F1] to use CALC.
- Press [F3] to select LinR. This puts LinR on the screen.
- You must now enter the names of the lists, L1 and L2 that contain the data. To do this, press [ALPHA] [7] [1][,] [ALPHA] [7][2].

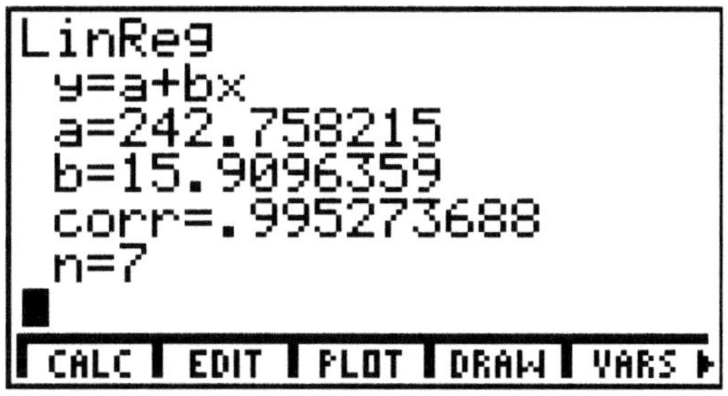

- Press [ENTER] to see the values of a and b as shown.
- Hence, the equation of the line through the data points is $y = 15.90963588x + 242.7582149$.
- Use the [2nd][MORE] keys to access MODE. Change line 2 to Float 3 to round these values to 3 decimals as shown.

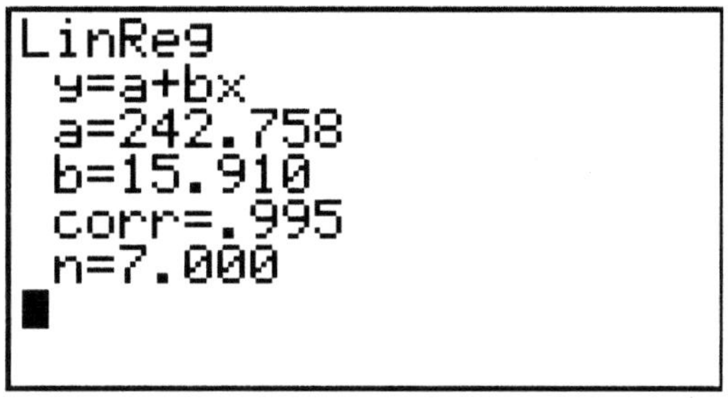

- Press [CLEAR] to return to the Home screen. Press [ENTER] to see the Linear Regression equation in 3 decimal places, $y = 15.910x + 242.758$.
- Note that the correlation coefficient is corr =.995.

Graph the Linear Regression Equation on the Graph Screen

We would like to graph the linear regression equation on the same screen as the scatter plot.

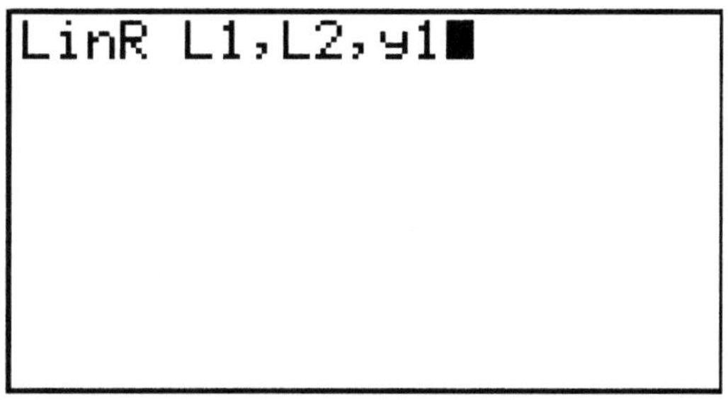

- Go to the graph menu. Clear all y= functions. Place the cursor on y1.
- Be sure that Plot 1 is on. Go back to the Home screen.
- Use [2nd][ENTER] to return to LinR L1, L2 screen. You ma y have to do this more than once. Immediately after L2 enter a comma and y1 using [,] [2nd] [ALPHA] [0] [1].

- Press [ENTER] to see the regression on the Home screen.
- Press the [GRAPH] key then [F1] to see the regression equation as y1 on the y= screen.

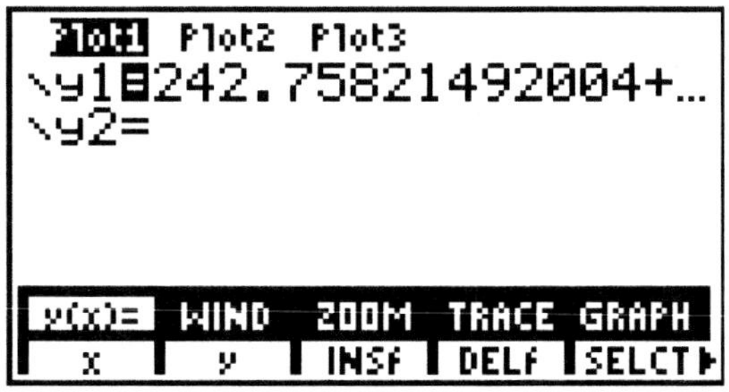

- Press [EXIT] [F5] to see the scatter plot and the linear regression equation graphed on the same screen.
- Press [F4] TRACE to trace the graphs.
- Use the up or down arrow to toggle between the scatter plot and the linear regression equation.

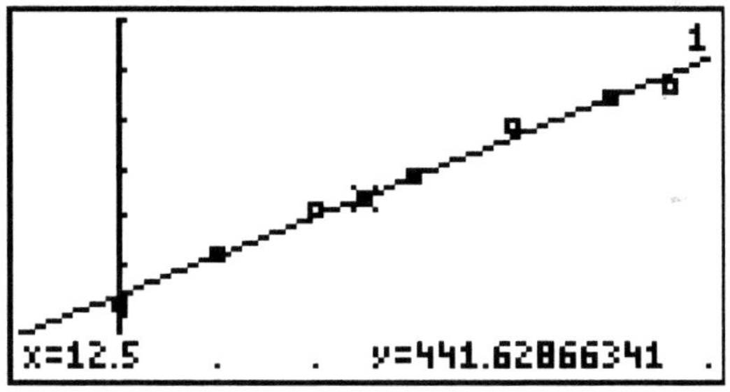

1.6 Example 1 using EXCEL

Example 1 of section 1.6 *Health Service Employment* gives a table of the numbers of full and part-time employees in offices and clinics of dentists in the selected years between 1970 and 1998. You are asked to create a scatter plot, graph some potential lines that may fit the data, and then find the linear regression line. Finally, you must graph the linear regression line and the scatter plot on the same grid.

Year	1970	1975	1980	1985	1990	1995	1998
x = # years past 1970	0	5	10	15	20	25	28
# Employees (in thousands)	222	331	415	480	580	644	666

Create the scatter plot - See 1.2 Example 5 for detailed instructions.

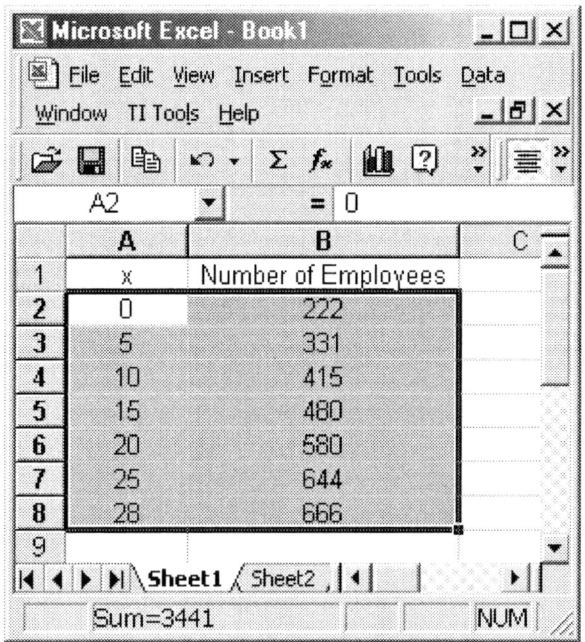

- Enter labels for two columns in EXCEL, namely x and the Number of Employees.
- Enter the values of x and the Number of Employees.
- Highlight the data set.
- Select **Chart Wizard** on the menu bar.
- Select *XY* Scatter, Scatter Compares pairs of values. Click on NEXT.
- Data range should be the range of cells you highlighted.
- Check Series in columns. Click on NEXT.
- Enter Titles, and *uncheck* major gridlines.
- Click on **Legends** tab and *uncheck* **Show Legend.** Click on **NEXT.**
- Select As object in.
- Click on **FINISH.** The scatter plot will appear in EXCEL.
- You may right click on the chart to format the chart at any time.

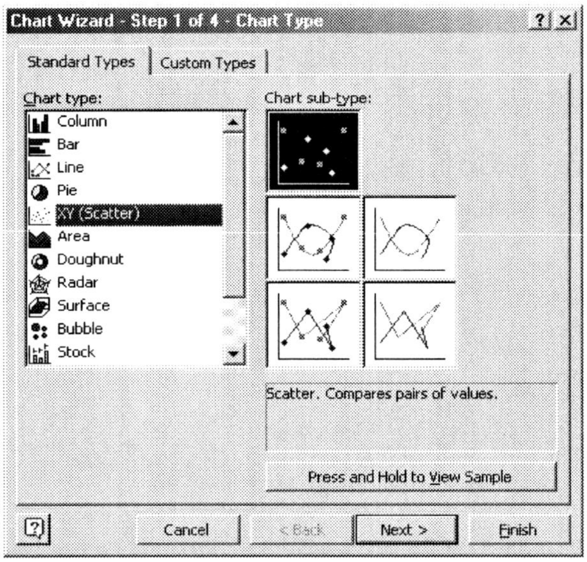
Chart Wizard - Step 1 of 4 - Chart Type
Standard Types
Custom Types
Chart type:
Column
Bar
Line
Pie
XY (Scatter)
Area
Doughnut
Radar
Surface
Bubble
Stock
Chart sub-type:
Scatter. Compares pairs of values.
Press and Hold to View Sample
Cancel
< Back
Next >
Finish

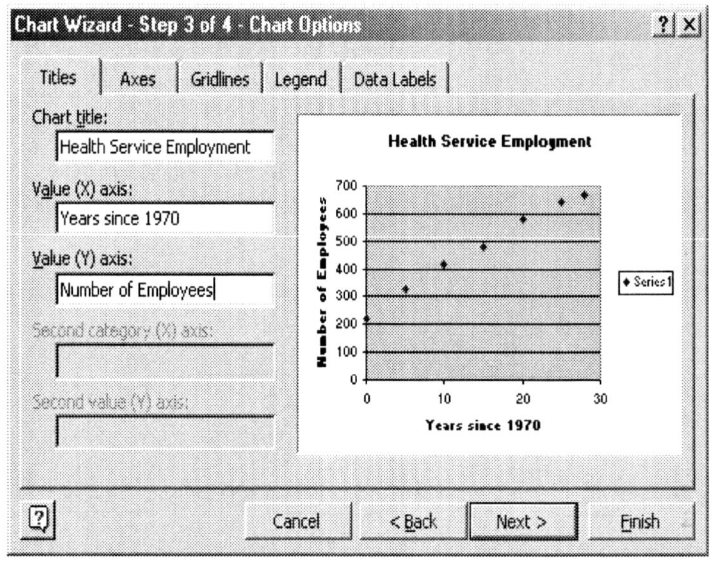
Chart Wizard - Step 3 of 4 - Chart Options
Titles
Axes
Gridlines
Legend
Data Labels
Chart title:
Health Service Employment
Value (X) axis:
Years since 1970
Value (Y) axis:
Number of Employees
Second category (X) axis:
Second value (Y) axis:
Health Service Employment
Number of Employees
Years since 1970
Series1
Cancel
< Back
Next >
Finish

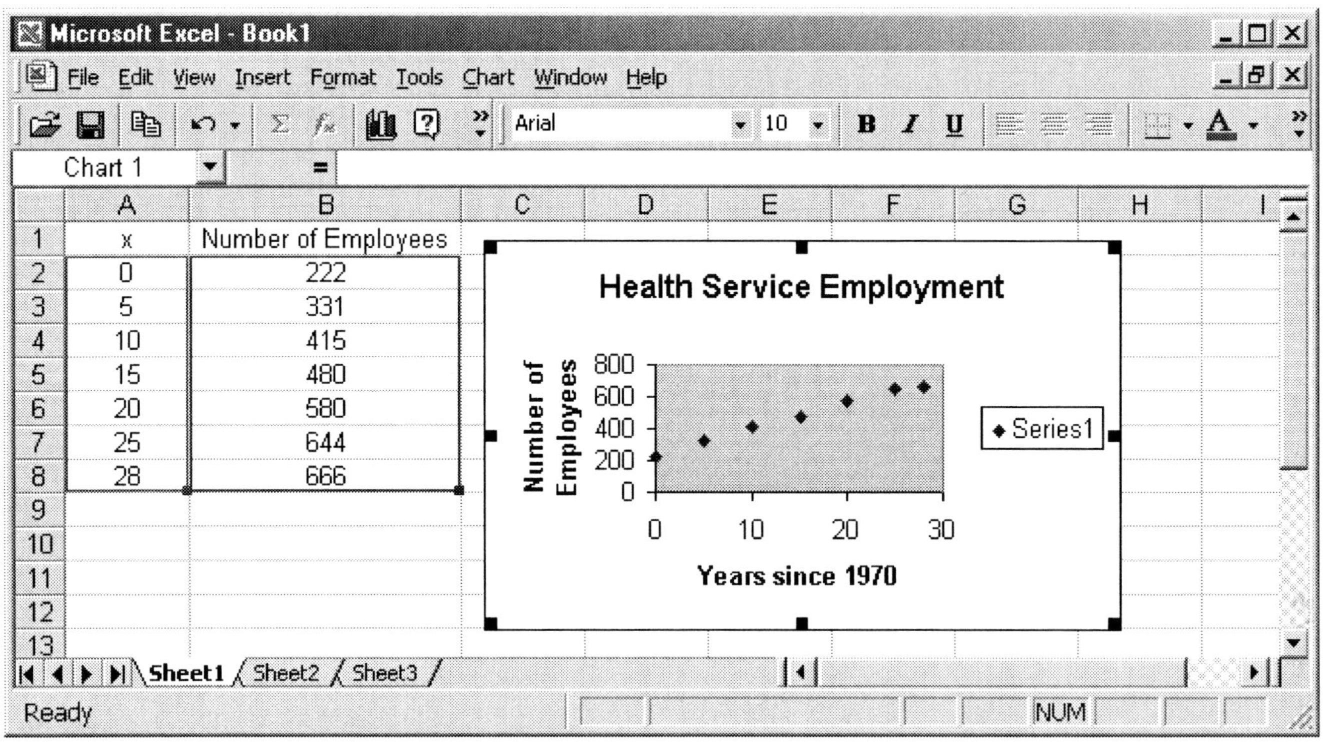
Microsoft Excel - Book1
File Edit View Insert Format Tools Chart Window Help
Arial
10
Chart 1
x
Number of Employees
0 222
5 331
10 415
15 480
20 580
25 644
28 666
Health Service Employment
Number of Employees
Years since 1970
Series1
Sheet1
Sheet2
Sheet3
Ready
NUM

Graph the three lines on the scatter plot

- Enter the three given equations into columns C, D, and E as shown.
- In cell C2 enter $= -12 * A2 + 660$, then press [ENTER]. Grab the right handle on C2 and drag.
- In cell D2 enter $= 13 * A2 + 220$, then press [ENTER]. Grab the right handle on D2 and drag.
- In cell E2 enter $= 16 * A2 + 242$, then press [ENTER]. Grab the right handle on E2 and drag.
- Highlight all 5 columns.
- Use Chart Wizard.
- Select Scatter with data points connected by smoothed lines.
- Enter Titles, etc. and click on **Finish** to see all three graphs.
- Use the mouse and left click on each line. The data points used to graph the line and the coordinates of the point are shown. By examination, it looks like y3 is the best fit for the original data.

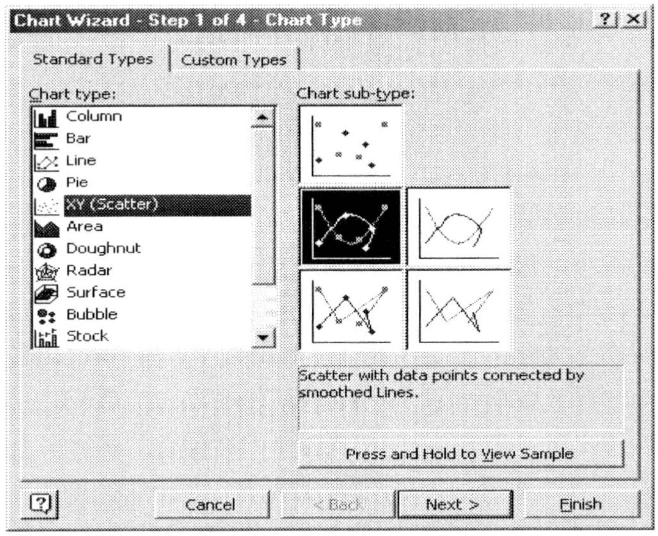

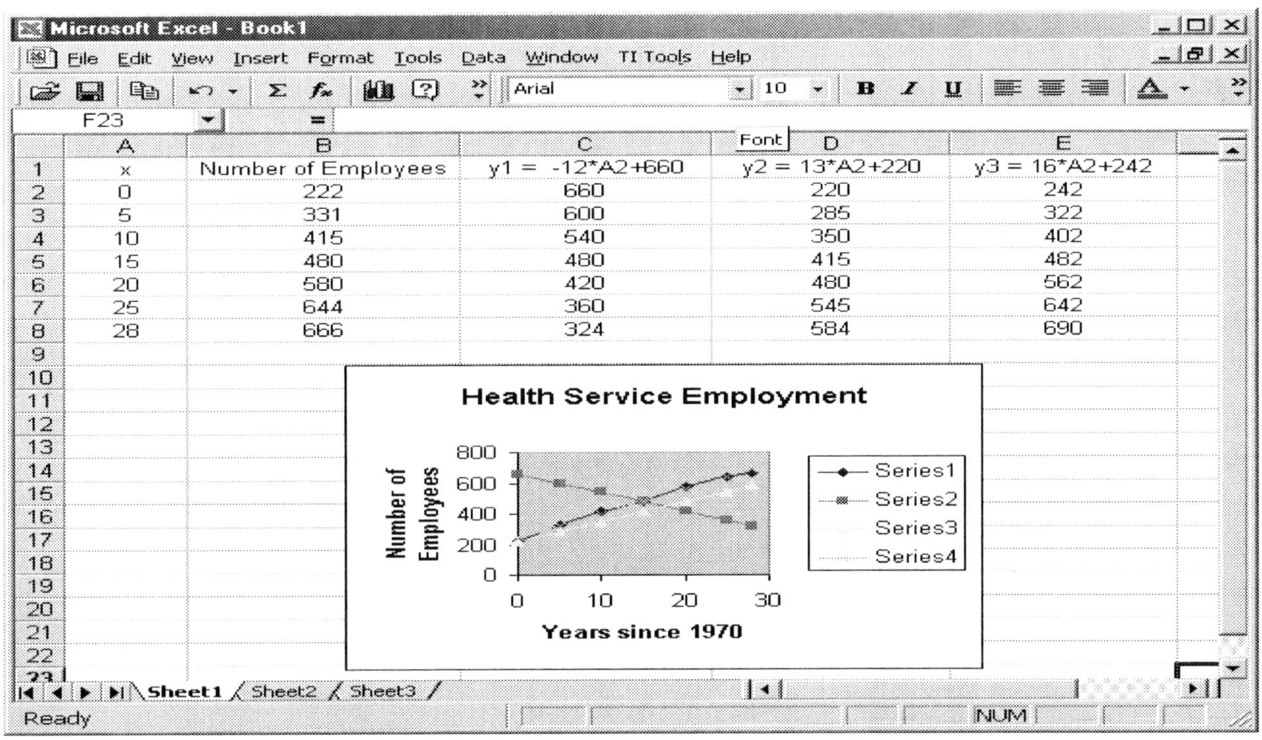

- To format the graph so that the original scatter plot is more prominent and the lines less prominent, as shown, locate one original data point of the scatter plot and double click on it. A **Format Data Series** dialog box appears as shown below.
- In the **Patterns** tab, select **None** under **Line**. Under **Marker** in the **Patterns** tab**,** go to **Custom** and select a prominent marker, like the dot.

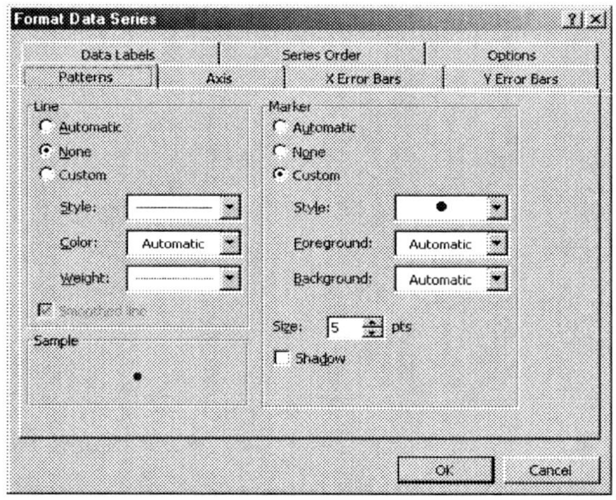

- Click **OK** to see the graph of the scatter plot and the sample lines.
- To make the lines less prominent, double click on one of the lines. In the **Patterns** tab, select **Custom** on the **Markers** side, and chose a marker that is less prominent, like the short dash, as shown.
- Repeat the previous step for the remaining two lines.
- Click anywhere off the graph to view the three lines and the scatter plot.
- Again, it looks like y3 is the best fit of the three sample lines.

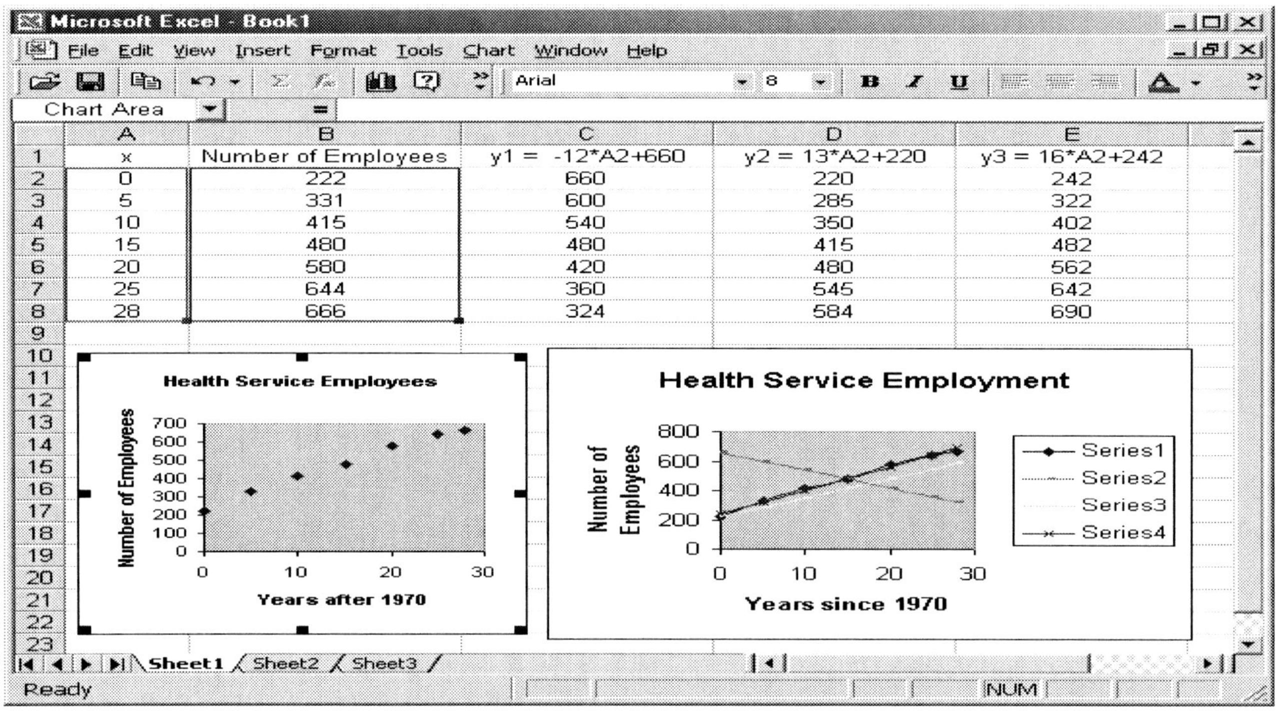

Find the regression equation and put it on the scatter plot.

- Click anywhere on the scatter plot. This puts **Chart** on the menu bar.
- Go to **Chart** on the menu bar.
- Select **Add Trendline**. Select **Linear**.

- Click on the Options tab and check Display equation on chart and Display r^2 on chart.

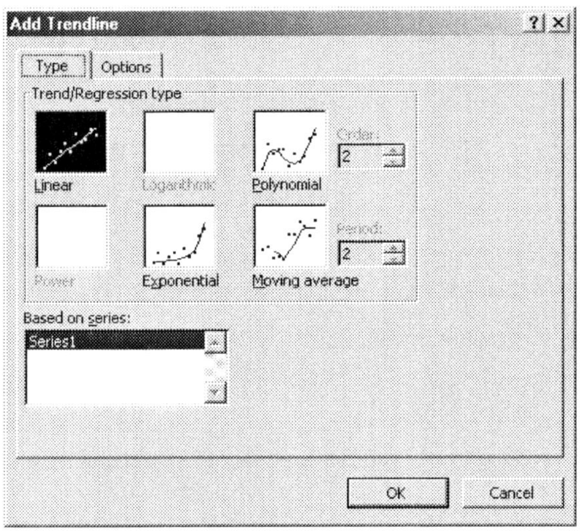

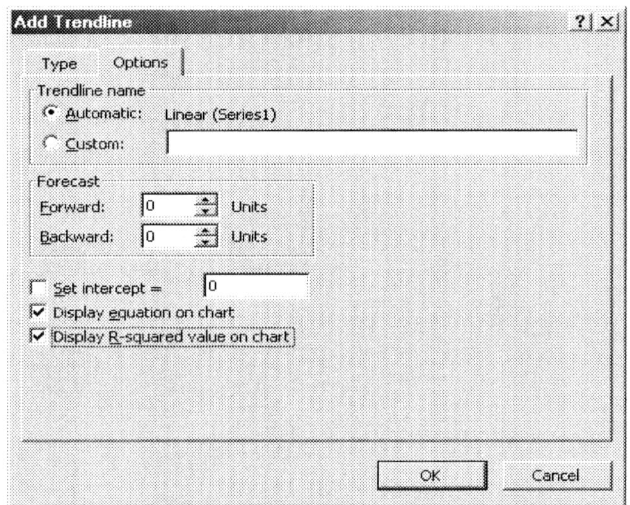

- Click **OK.**
- You can grab and move the equation box anywhere you wish on the chart.
- Click **OK.** The linear regression line is drawn on the graph, and its equation and the value of r^2 are also shown.
- You may wish to move the box that contains the equation to a different location. To do this, left click on the equation and hold it down while you move the box to a new location.

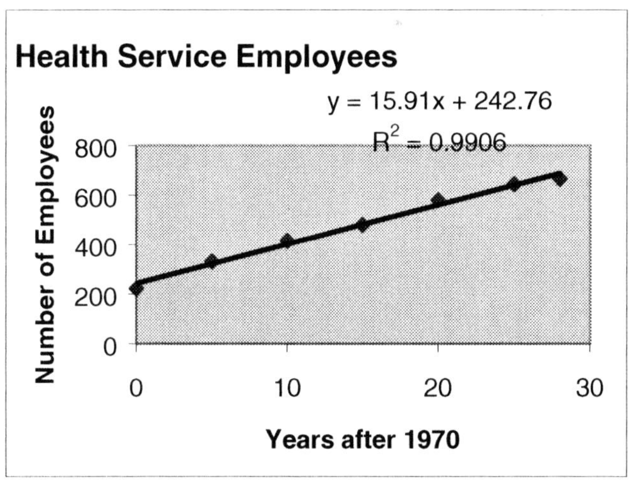

- Take the square root of $\mathbf{r^2 = .9906}$ to find the correlation coefficient, $\mathbf{r = .995}$.

1.7 Example 2 on the TI-83, TI-83 Plus, and TI-84 PLUS

This example asks you to solve a system of equations by graphing. You can use the intersection feature of the graphing calculator to do this. See Section 1.5 Example 5 for more detailed instructions on finding an intersection.

Solve a system of linear equations by graphing.

Solve the system $\begin{cases} 3x - 4y = 21 \\ 2x + 5y = -9 \end{cases}$ by graphing.

Find the exact coordinates of the Intersection point

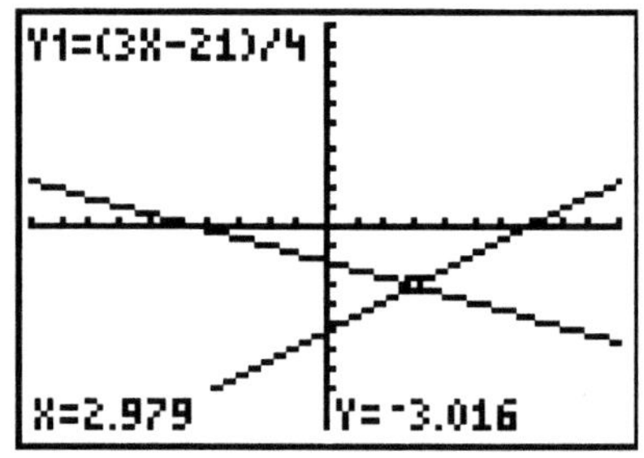

- Turn off or clear any equations or plots.
- Solve each equation for *y* and enter them as Y1 and Y2. Enter Y1=$(3x-21)/4$ and Y2=$(-9-2x)/5$.
- Press the [ZOOM] key. Select 6:Zstandard to set the window to [-10,10] by [-10, 10].
- Press the [GRAPH] key and then [TRACE] to see the graph. Use the arrow to move the cursor close to the intersection point.

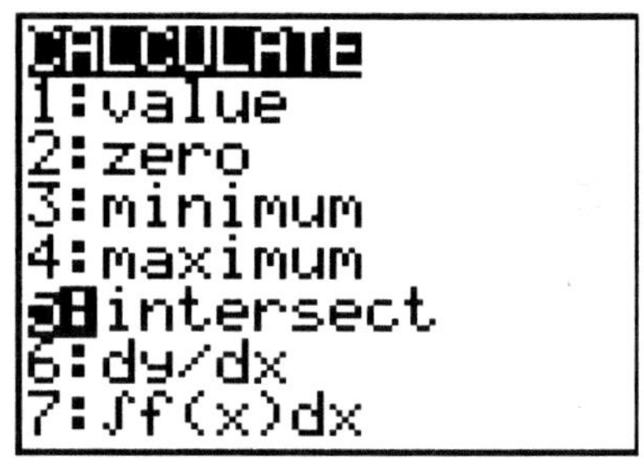

- Press [2nd] [TRACE] to access CALC. A CALCULATE screen appears.
- Press the [5] key or cursor down to 5:intersect and press [ENTER].
- You are brought back to the graph screen.
- You must select a *First curve*. Press [ENTER].
- Next you must select a *Second curve*. Use the up or down arrow key to move to the other line. Press [ENTER].

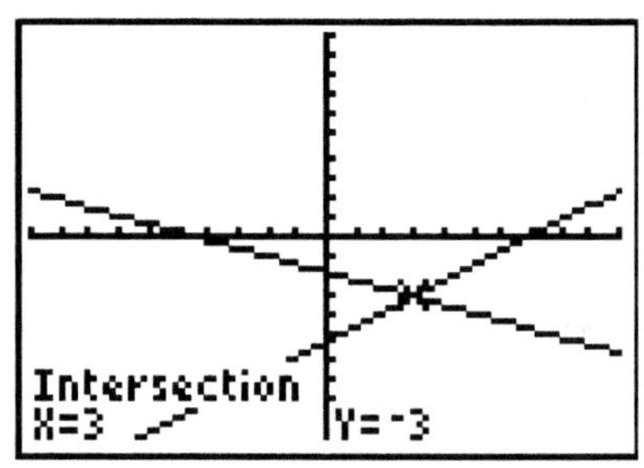

- Next, you must enter a *Guess*. You can enter a value somewhere within the interval defined by your left and right bounds. However, you can just press [ENTER].
- Finally, press [ENTER] to see the coordinates of the Intersection at the bottom of the screen.

1.7 Example 2 on the TI-86

This example asks you to solve a system of equations by graphing. You can use the intersection feature of the graphing calculator to do this. See Section 1.5 Example 5 for more detailed instructions on finding an intersection.

Solve a system of linear equations by graphing.

Solve the system $\begin{cases} 3x-4y=21 \\ 2x+5y=-9 \end{cases}$ by graphing.

Find the exact coordinates of the Intersection point

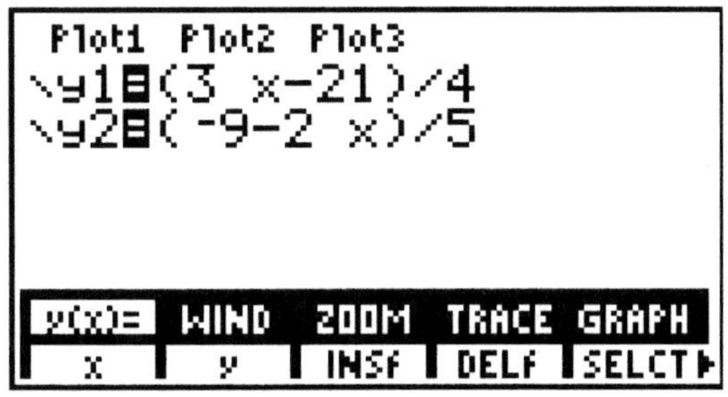

- Turn off or clear any equations or plots.
- Solve each equation for y and enter them as y1 and y2. Enter $y1=(3x-21)/4$ and $y2=(-9-2x)/5$.
- Set the window to [-10, 10] by [-10, 10]. Use xScl = 1 and yScl = 1.
- Press the [EXIT] [F5] keys to select GRAPH.
- Press [F4] TRACE to walk along the curves. Use the arrow to move the cursor close to the intersection point.

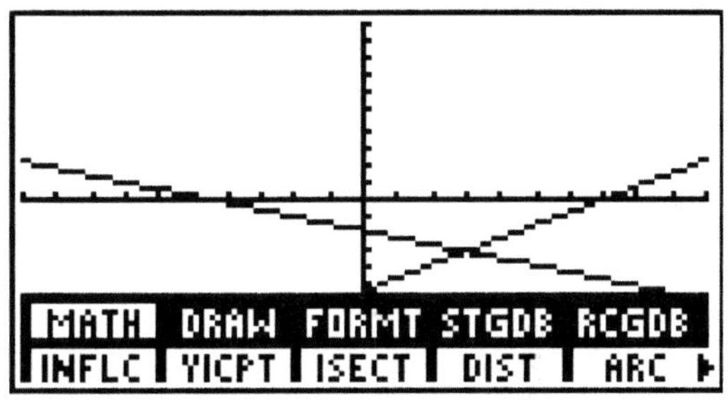

- Press [EXIT] to see the Graph menu again.
- Press [MORE] [F1] to access the MATH menu of the graph screen.
- Press the [MORE][F3] key to select ISECT. This will locate the intersection point of the two lines.
- You must select the *First curve* to tell the calculator which line you want to use first. The number in the upper right corner of the screen tells you which curve you are on. The 1 says you are on y1 and the 2 says you are on y2. Press [ENTER].

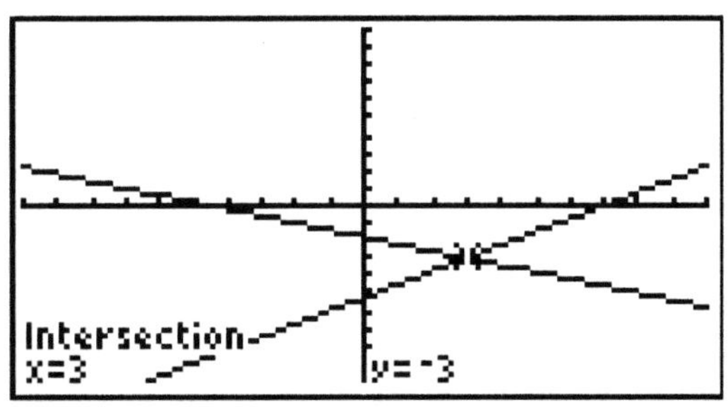

- Next you must select a *Second curve* to tell the calculator which second curve to use for the intersection. Notice that the cursor is now on the other line. Press [ENTER].
- Next, you must enter a *Guess*. You can enter a value or you can just press [ENTER].
- Finally, you see the coordinates of the Intersection at the bottom of the screen.

1.7 Example 2 using EXCEL

This example asks you to solve a system of equations by graphing. You can use the intersection feature of the graphing calculator to do this. See Section 1.5 Example 5 for more detailed instructions on finding an intersection.

Solve a system of linear equations by graphing.

Solve the system $\begin{cases} 3x-4y=21 \\ 2x+5y=-9 \end{cases}$ by graphing.

- Solve each equation for *y*.Enter some *x*- values in row 1.
- Click on cell **B1** and enter the first equation of the system as $=(3*A1-21)/4$. Grab the right handle and drag the equation through the columns of the values of *x*.
- Click on cell **C1** and enter the second equation of the system as $=(-9-2*A1)/5$. Grab the right handle and drag the equation through the columns of the values of *x*.
- Select the 3 rows and columns that contain the data.
- Go to Chart Wizard and select *X-Y* Scatter. Select the scatter plot connected by lines without markers. Select **Next.**
- Select Data in **series in rows** because the data was entered in rows. Note that this could also be done with columns by putting the *x*-values into column A, the *y*1 equation into column B, and the *y*2 equation into column C. Scroll through the *x*-values used. If you do this, then select series in columns.
- Click **Next.** Enter the Titles, uncheck major gridlines, and do not show series box.
- Click **Next** to put the graph in the worksheet. Click **Finish.**

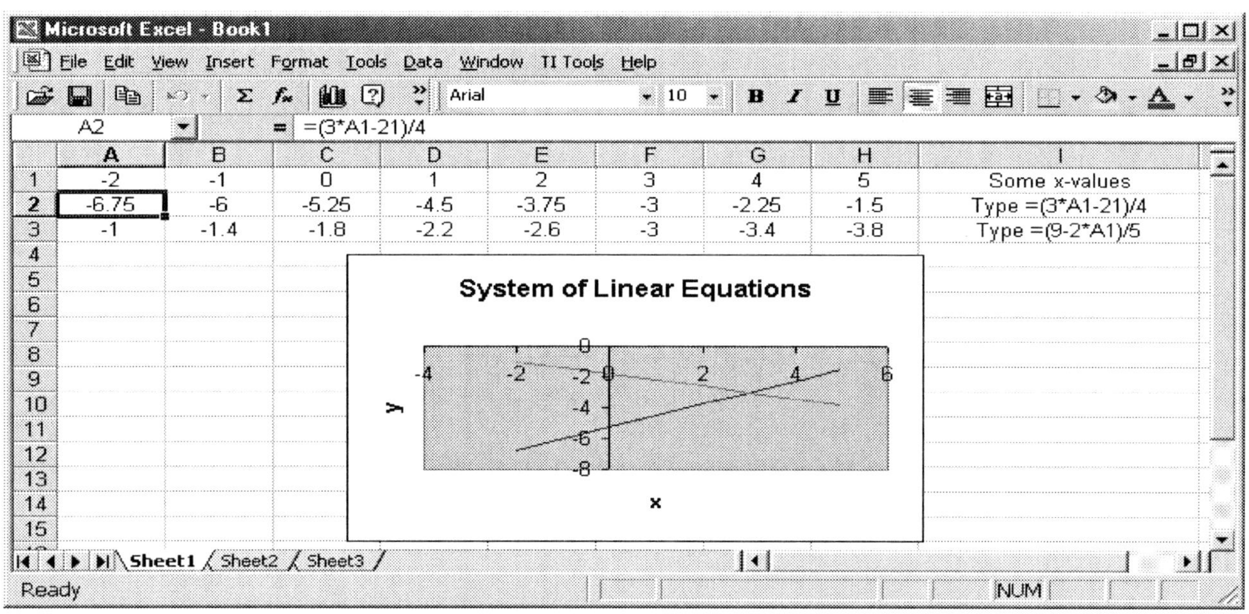

Find the exact coordinates of the Intersection point using GOAL SEEK in EXCEL.

To use Goal Seek in EXCEL to find the intersection of two functions, you must first find the difference of the two functions $0=(3x-21)/4-(-9-2x)/5$). This will give you the x-coordinate of the intersection point. You can go back to the data points and use this value as one of the x's.

- First enter a label for the first two columns, say x in cell A22 and the difference of the linear equations in cell B22 as $y_1 - y_2$.
- Enter any value of x, say 1, into cell A23 and the equation as $=((3*A23-21)/4)-(-9-2*A23)/5$ into cell B23. Note that you can click on cell A23 to put it in the formula.
- Go to **Tools** on the toolbar. Select **Goal Seek.** A dialog box appears. Fill in the dialog box as shown.
- Another dialog box appears that tells you a solution has been found. Click **OK.**
- Cell A23 now contains the value 3 and cell B23 contains 0. The solution is that the x-coordinate of the intersection is 3. To find the y-value, go back to the data set and notice that $y=-3$ when $x=3$ for both of the functions in the system.

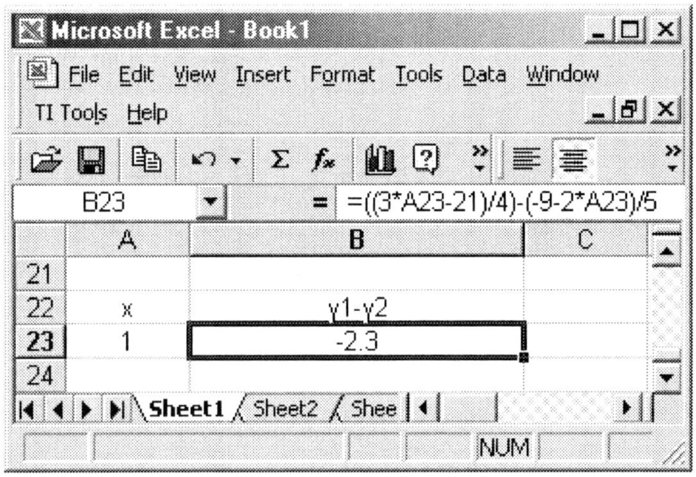

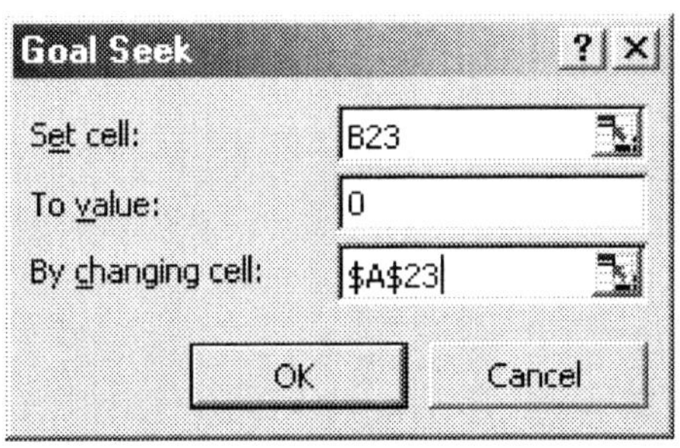

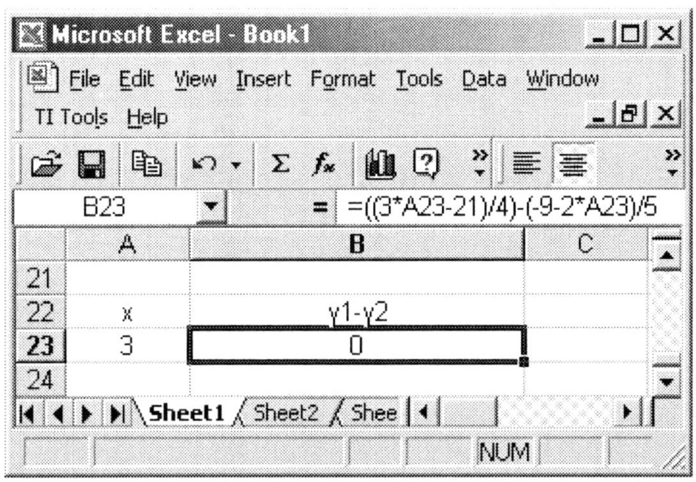

CHAPTER 2

QUADRATIC AND OTHER NONLINEAR FUNCTONS

2.1 Example 1 on the TI-83, TI-83 PLUS, and TI-84 PLUS

Vertices of Parabolas

To find the vertex of a parabola, first graph the parabola, use the formula $x = \frac{-b}{2a}$ to find the x-coordinate of the vertex, then find the y-value of the function at the vertex to determine the appropriate window for the function.

Example 1 asks you to find the vertex of $P(x) = -2x^2 + 60x$.

Display the vertex on the Graph of the function $P(x) = -2x^2 + 60x$

- Press the [Y=] key and enter Y1 $= -2x^2 + 60x$.
- Press [GRAPH] to see the graph.
- Using the formula $x = \frac{-b}{2a}$ you find that the x-coordinate of the vertex is 15.
- Set the window to include the value $x = 15$.
- Press the [TRACE] key.
- At the bottom of the screen, type the number 15. Press [ENTER].
- The y-value of the vertex, $y = 450$, appears on the bottom of the screen.
- Set the window accordingly to [0,30,5] by [0,500,50].
- Press [GRAPH] to view the graph.

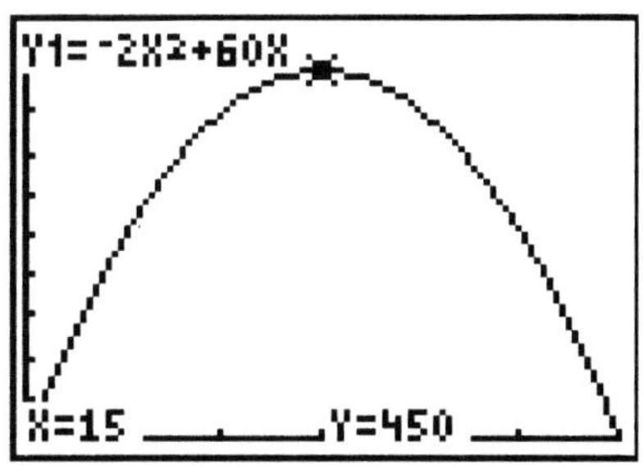

2.1 Example 1 on the TI-86

Vertices of Parabolas

To find the vertex of a parabola, first graph the parabola, use the formula $x = \frac{-b}{2a}$ to find the x-coordinate of the vertex, then find the y-value of the function at the vertex to determine the appropriate window for the function.

Example 1 asks you to find the vertex of $P(x) = -2x^2 + 60x$.

Display the vertex on the Graph of the function $P(x) = -2x^2 + 60x$

- Press the [GRAPH] key, then [F1] to enter y1 $= -2x^2 + 60x$.
- Press [EXIT] [F5] to see the graph.
- Using the formula $x = \frac{-b}{2a}$ you find that the x-coordinate of the vertex is 15.
- Press the [TRACE] key.
- Set the window to include the value $x = 15$.
- At the bottom of the screen, type the number 15. Press [ENTER].
- The y-value of the vertex appears on the bottom of the screen.
- Set the window accordingly to [0,30,5] by [0,500,50].
- Press [GRAPH] to view the graph.

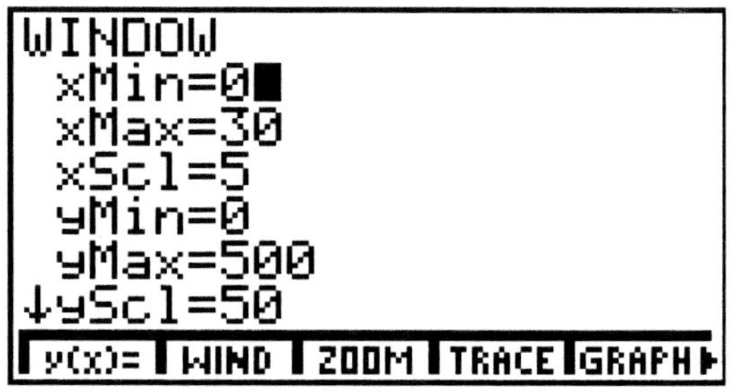

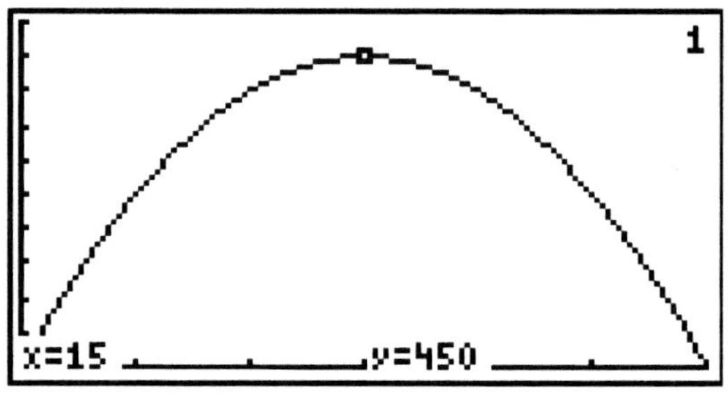

2.2 Example 5 on the TI-83, TI-83 PLUS, and TI-84 PLUS

Marijuana Use

Example 5 uses the intersection method of graphing and a table to find when the percent of high school seniors who have tried marijuana predicted by the function $p = -0.2286t^2 + 2.7783t + 49.8138$ reached 33%, where t is the number of years after 1975.

Use the Intersection Method to solve $33 = -0.2286t^2 + 2.7783t + 49.8138$ - See 1.5 Example 5 for more details of this method

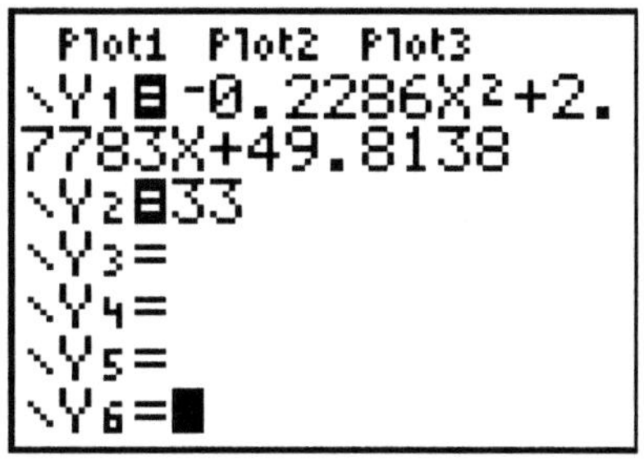

- Press the [Y=] key and enter Y1 $= -0.2286x^2 + 2.7783x + 49.8138$ and Y2 = 33.
- Set the window to [0,20,5] by [0,100,10].
- Press [GRAPH] to see the graphs.
- Move the cursor close to the intersection point of Y1 and Y2.
- Press [2nd] [TRACE] (CALC), select 5:intersect.
- Press [ENTER] to select the first curve.
- Press [ENTER] to select the second curve.
- Press [ENTER] to Guess.
- The coordinates of the intersection point, appear at the bottom of the screen.

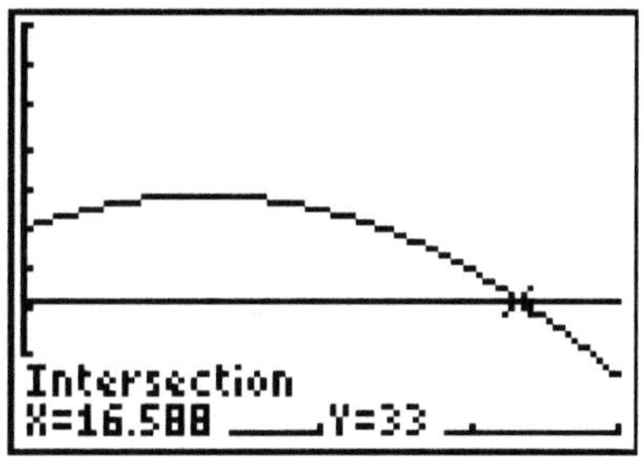

Use a Table to find the solution

- Press [2nd] [WINDOW] (TBLSET) to set the table as shown.
- Press [2nd] [GRAPH] (TABLE) to view the table. The solution occurs between 16 and 17 years.

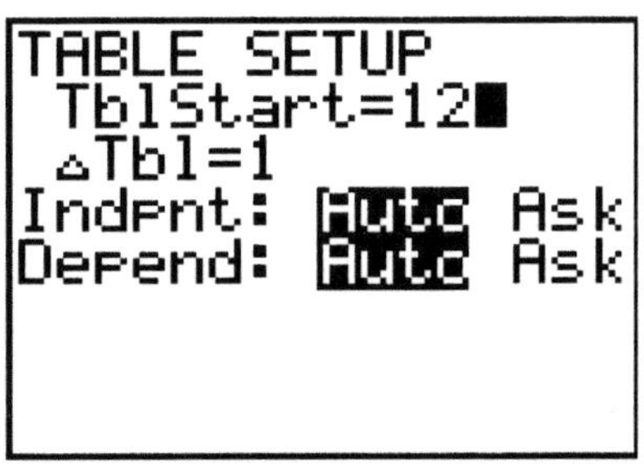

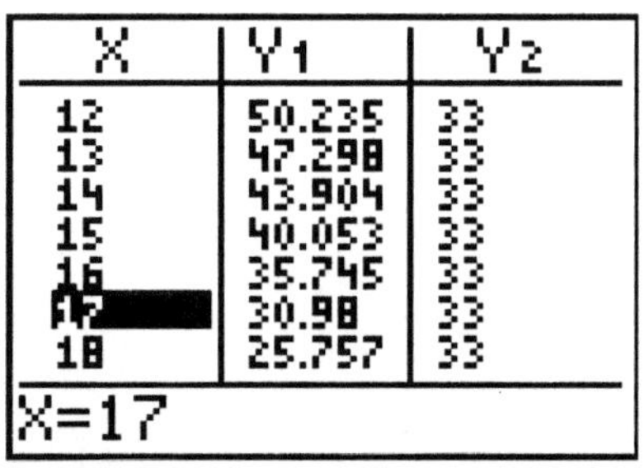

2.2 Example 5 on the TI-86

Marijuana Use

Example 5 uses the intersection method of graphing and a table to find when the percent of high school seniors who have tried marijuana predicted by the function $p = -0.2286t^2 + 2.7783t + 49.8138$ reached 33%, where t is the number of years after 1975.

Use the Intersection Method to solve $33 = -0.2286t^2 + 2.7783t + 49.8138$
- See 1.5 Example 5 for more details of this method

- Press the [GRAPH] [F1] and enter y1 $= -0.2286x^2 + 2.7783x + 49.8138$ and y2 $= 33$.
- Set the window to [0,20,5] by [0,100,10].
- Press [F5] to see the graphs.
- Move the cursor close to the intersection point of y1 and y2.
- Press [EXIT] [MORE][F1] to access the MATH menu.
- Press [MORE] [F3] to select ISECT.
- Press [ENTER] to select the first curve.
- Press [ENTER] to select the second curve.
- Press [ENTER] to Guess.
- The coordinates of the intersection point, appear at the bottom of the screen.

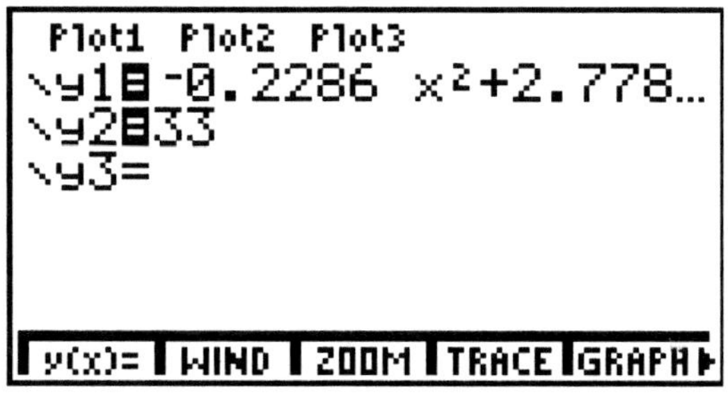

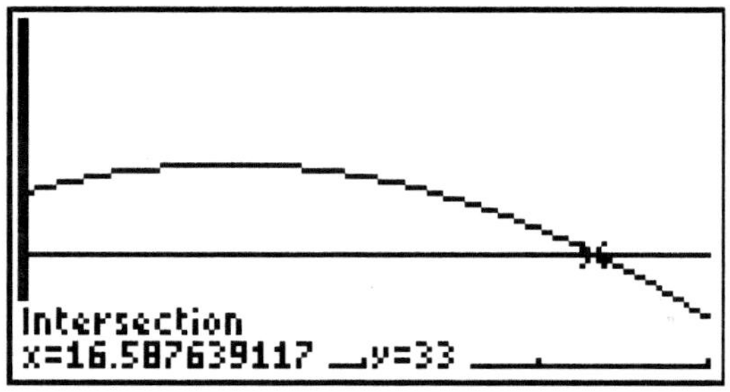

Use a Table to find the solution

- Press the [TABLE] to set the table as shown.
- Press [F1] (TABLE) to view the table. The solution occurs between 16 and 17 years.

2.3 Graphing More Nonlinear Functions on the TI-83, TI-83 and TI-84 PLUS

This section introduces several nonlinear functions and their graphs.

Power Functions such as $C = 0.11W^{1.54}$ of Example 1

- Set the window to [0,20,5] by [0,10,1].
- Press the [Y=] key.
- Enter the function as $Y1 = 0.11x \wedge 1.54$.
- Press the [GRAPH] key to see the graph.

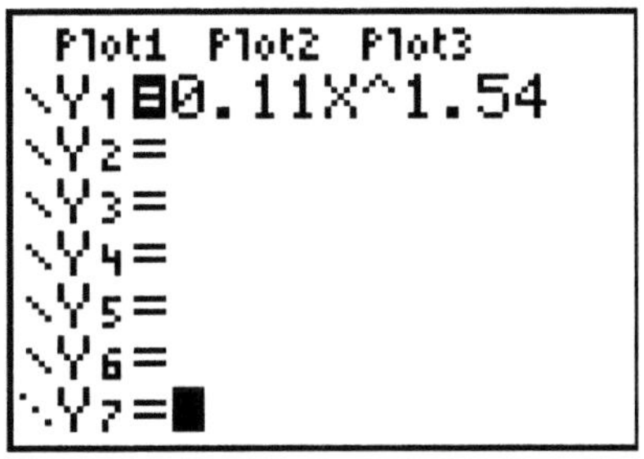

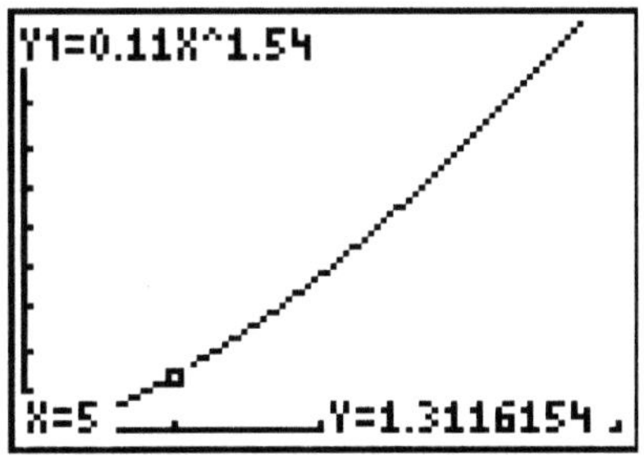

Root Functions such as $y = x^{2/3}$

There are two ways to enter the root functions, one uses the caret ^ key, and the other uses the $\sqrt[x]{\ }$ in the Math menu.

Using the caret key ^

- Set the standard window.
- Press the [Y=] key.
- Enter the function as $Y1 = x \wedge (2/3)$. Note that the parentheses around 2/3 are essential.
- Press the [GRAPH] key to see the graph.

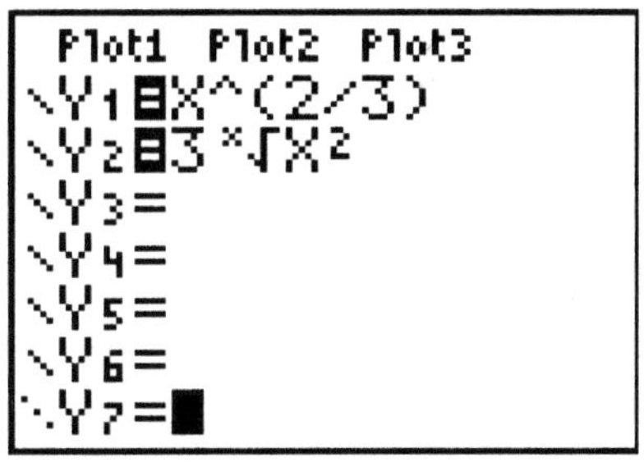

Using $\sqrt[x]{\ }$ in the Math Menu

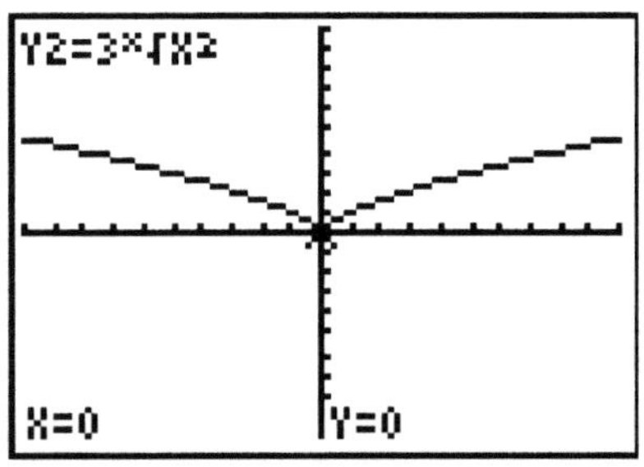

- Press the [Y=] key.
- Enter the function as $Y2 = 3\sqrt[x]{\ }x^2$. To do this, first type the 3, then press the [MATH] key and select 5: $\sqrt[x]{\ }$.
- Press the [GRAPH] key to see the graph.

Reciprocal Functions Graph $y = \dfrac{1}{x+5}$

Change the MODE to DOT

- Press the MODE key.
- Scroll down to the line CONNECTED DOT.
- Move the cursor over the word DOT and press ENTER.
- The graphs will be shown as a series of dots rather than connected line segments.

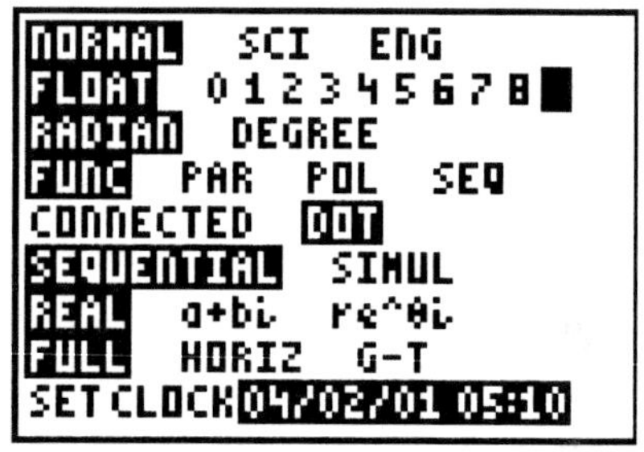

Graph the function

- Press the Y= key.
- Set the window as shown.
- Enter the function as $\text{Y1} = 1/(x+5)$.
- Note that the parentheses in the denominator are essential.
- Press the GRAPH key to see the graph.

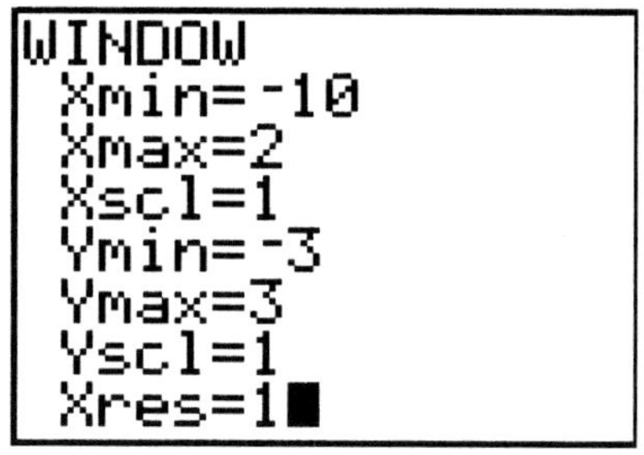

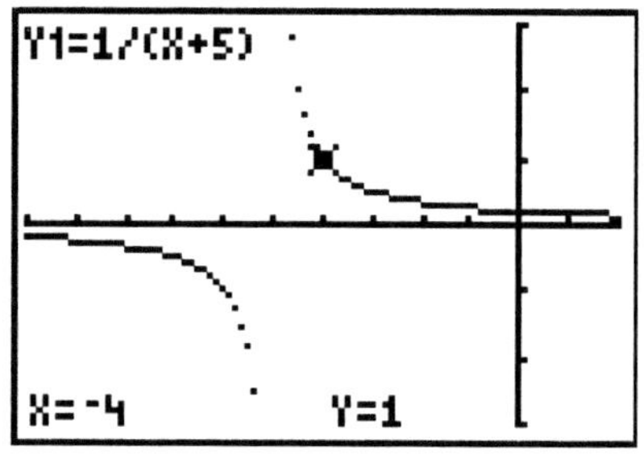

Graph a Piece-Wise Function $f(x) = \begin{cases} 5x+2 & \text{if } 0 \le x < 3 \\ x^3 & \text{if } 3 \le x \le 5 \end{cases}$ of

Example 3

- Set the window as shown.
- Enter the first piece of the function as $\text{Y1} = (5x+2)(0 \le x)(x < 3)$. The inequalities are found in the TEST menu. Press 2nd MATH (TEST). Select the appropriate inequality.
- Enter $\text{Y2} = (x \wedge 3)(3 \le x)(x \le 5)$.

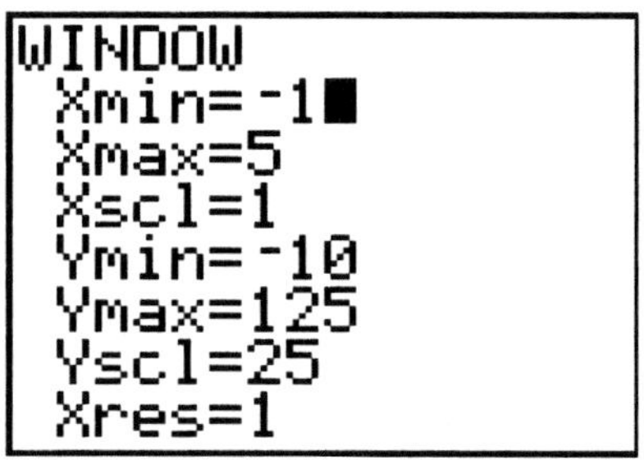

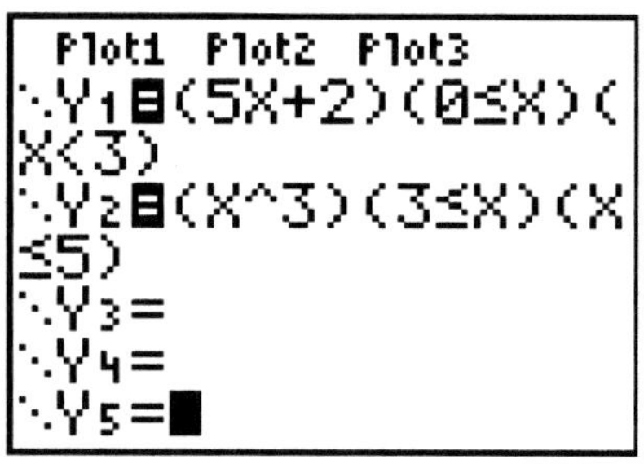

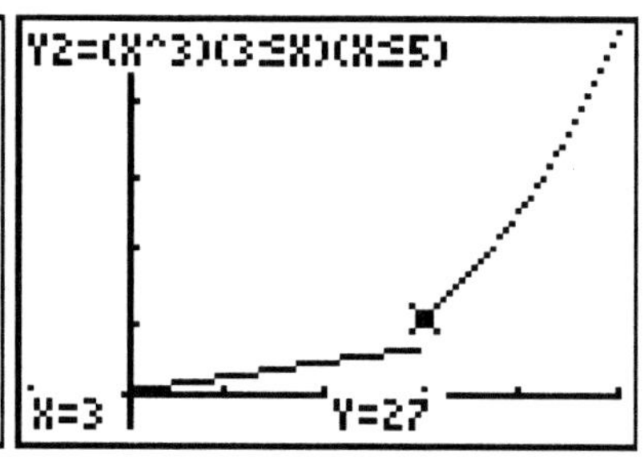

Absolute Value Function $|x| = \begin{cases} x & if\ x \geq 0 \\ -x & if\ x < 0 \end{cases}$

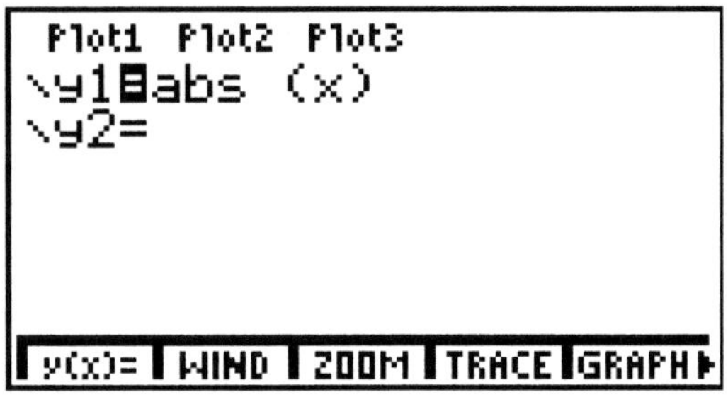

- Set the standard window.
- Place the cursor on y1=. The absolute value function is found in the Math menu as *abs*.
- Press the [2nd][×] (MATH) to access the Math menu.
- Press [F1] to select the NUM menu.
- Press [F5] to select *abs*.
- Type the argument, x, in parentheses to see y1 $= abs(x)$.
- Press the [EXIT] [EXIT] [F5] key to see the graph.

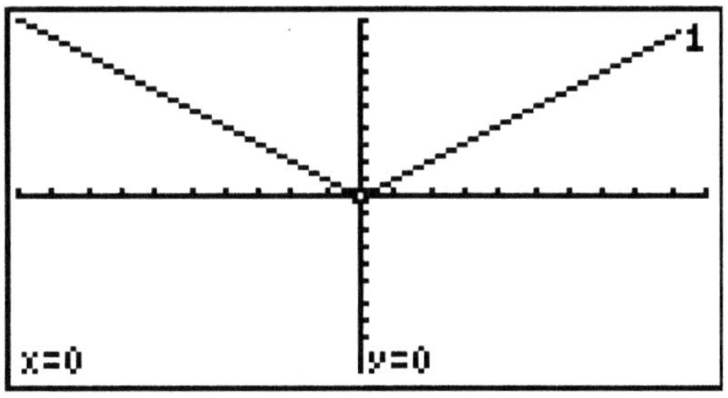

2.3 Graphing More Nonlinear Functions on the TI-86

This section introduces several nonlinear functions and their graphs.

Power Functions such as $C = 0.11W^{1.54}$ of Example 1

- Set the window to [0,20,5] by [0,10,1].
- Press the [GRAPH] key and press [F1] to enter the function.
- Enter the function as $y1 = 0.11x \wedge 1.54$.
- Press [EXIT] [F5] (GRAPH) to see the graph.

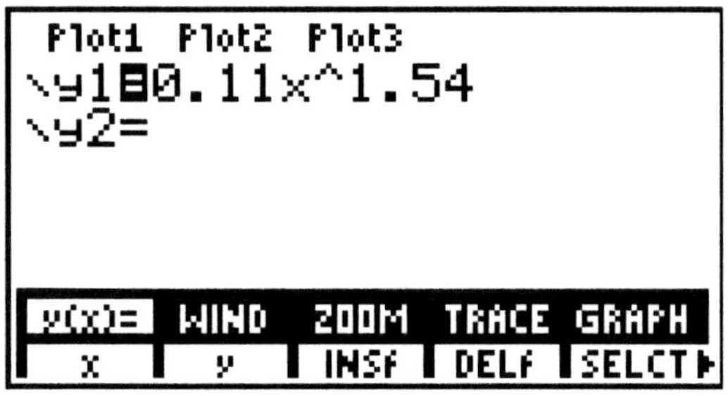

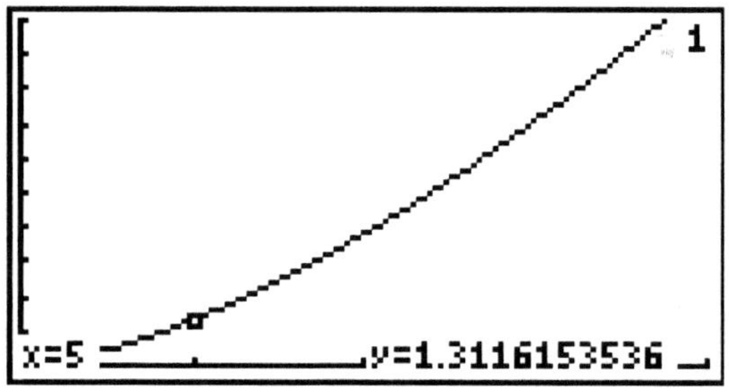

Root Functions such as $y = x^{2/3}$

There are two ways to enter the root functions, one uses the caret ^ key, and the other uses the $\sqrt[x]{\ }$ in the Math menu.

Using the caret key ^

- Set the standard window. Press the [GRAPH] key and press [F1] to enter the function.
- Enter the function as $Y1 = x \wedge (2/3)$. Note that the parentheses around 2/3 are essential.
- Press [EXIT] [F5] (GRAPH) to see the graph.

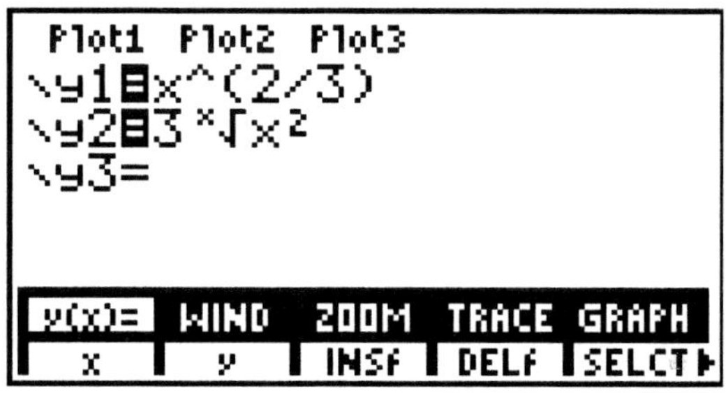

Using $\sqrt[x]{\ }$ in the Math Menu

- Press the [GRAPH] key and press [F1] to enter the function Enter the function as $Y2 = 3\sqrt[x]{\ }x^2$. To do this, first type the 3, then press the [2nd] [×] (MATH) to access the Math menu.
- Press [F5] (MISC) then [MORE] [F4] to put $\sqrt[x]{\ }$ on the screen. Then type x^2.
- Press [EXIT] [F5] (GRAPH) to see the graph.

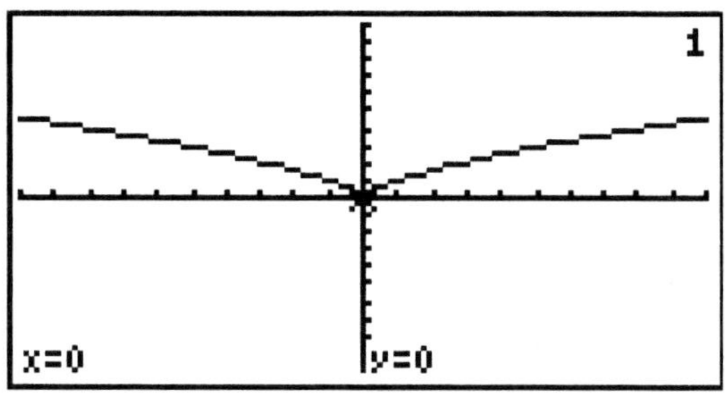

Reciprocal Functions Graph $y = \dfrac{1}{x+5}$

Format the Graph to DrawDot

- Press the [GRAPH] key then the [MORE] key, then [F3] (FORMT).
- Scroll down to the line DrawLine DrawDot.
- Move the cursor over the word DrawDot and press [ENTER].
- The graphs will be shown as a series of dots rather than connected line segments.

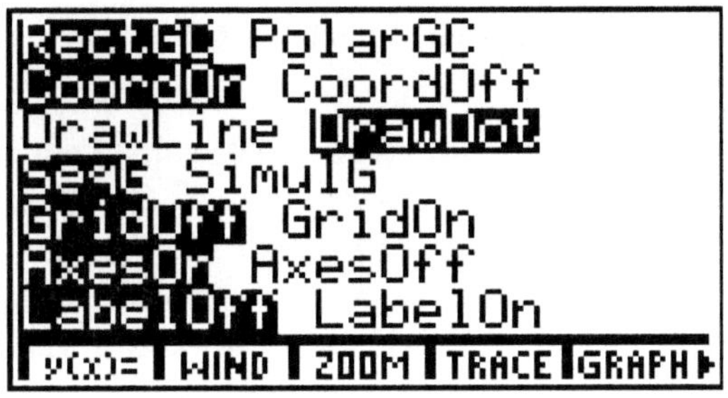

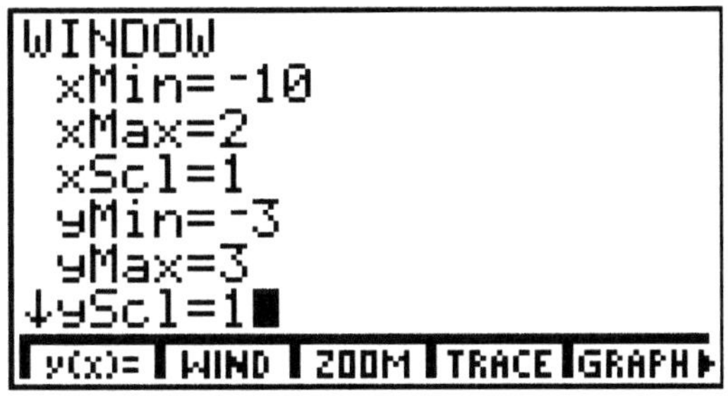

Graph the function

- Set the window as shown.
- Press the [F1] key to enter the function.
- Enter the function as $Y1 = 1/(x+5)$.
- Note that the parentheses in the denominator are essential.
- Press the [GRAPH] key to see the graph.

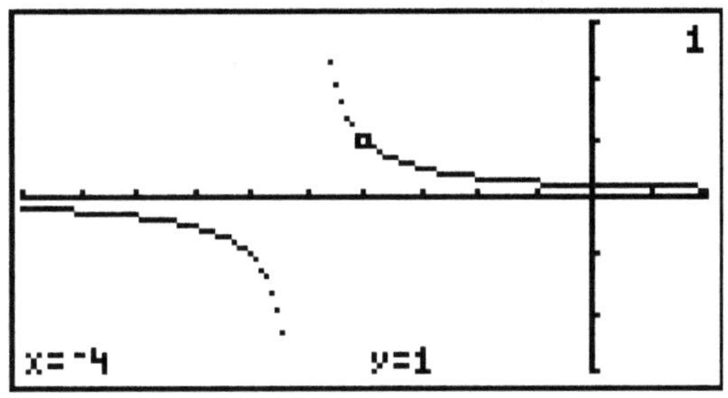

2.5 Example 1 on the TI-83, TI-83 Plus, and TI-84 PLUS

This example asks you to create a quadratic function that models the number of registered cocaine addicts in the United Kingdom for the years 1985-1995, using the number of years past 1985 as the input x.

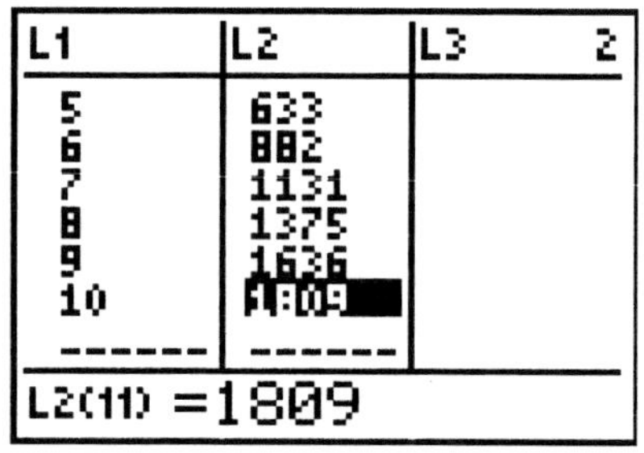

Create the scatter plot - See 1.2 Example 5 for detailed instructions.

- Set the window to [-1, 15, 1] by [300, 1900, 100] or use [ZOOM] 9: ZoomStat to set the window for the data. The data will be displayed over two screens. The second screen is shown.
- Use [STAT] EDIT 1: Edit to enter the x values 0 thru 10 into L1 and the number of registered cocaine addicts into L2. Reset the Mode to the word Float.
- Clear the lists if needed. Cursor to the name of the list to be cleared, press [CLEAR] [ENTER].
- Press [2nd] [Y=] to turn on and format Plot 1. Be sure that the data is coming from L1 and L2.
- Press the [GRAPH] key and then [TRACE] to see the scatter plot. Use the right and left arrows to move from point to point.

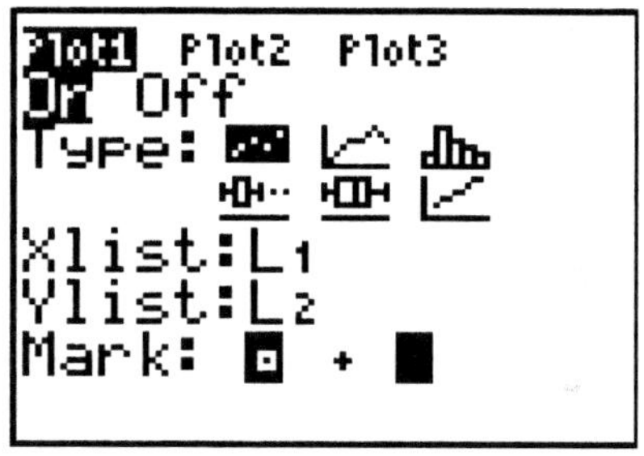

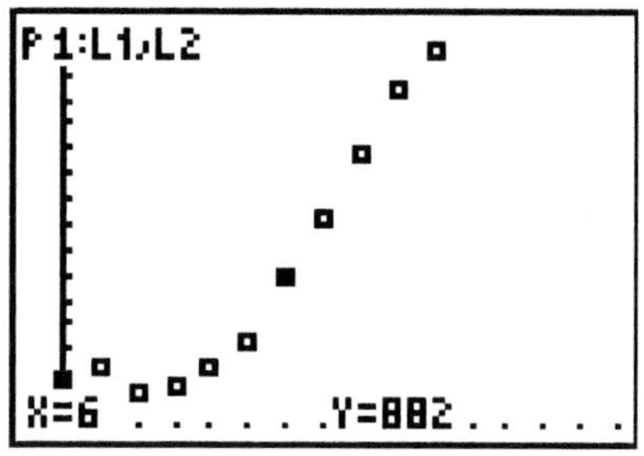

Find the Quadratic Equation to Model the Data.

- The data is already entered into L1 and L2, and Plot 1 is formatted for scatter plot.
- Reset the Mode to Float 3. Otherwise, you can round by hand.
- Next, press the [STAT] key and use the right arrow to go to CALC.
- Select 5:QuadReg. This puts QuadReg on the Home screen.
- Press [ENTER] to see the values of a , b and c as shown.
- Hence, the equation of the quadratic model through the data points is $y = 19.867x^2 - 56.999x + 489.280$.
- Turn DiagnosticOn if you do not see r^2.

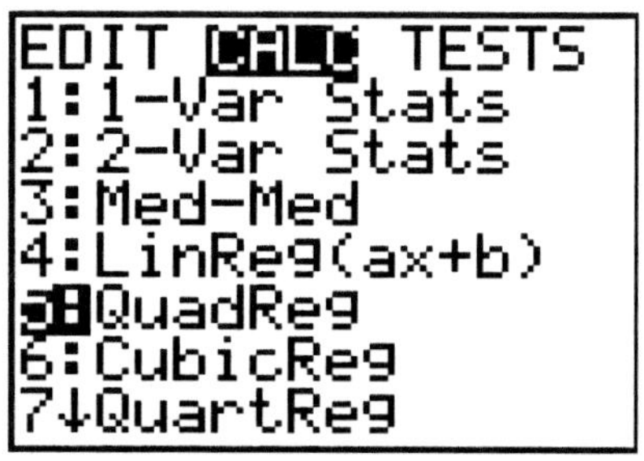

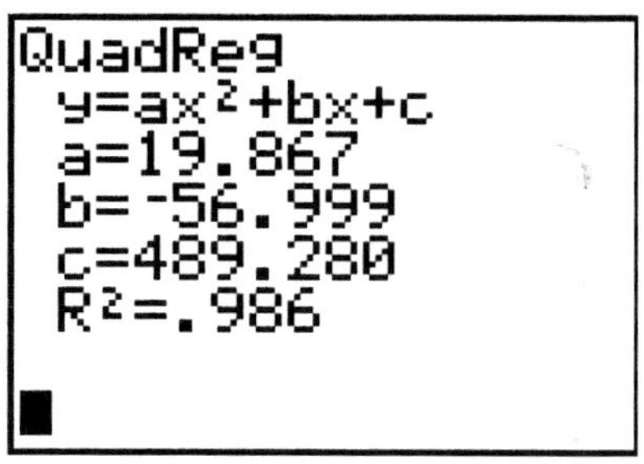

Graph the Quadratic Regression Equation on the Graph Screen

- Turn off all Y= functions. Be sure that Plot 1 is on.
- Move the cursor to the Y= line you wish to use.
- Press the [VARS] key. Select 5:Statistics.
- Move the right arrow twice to see the EQ menu.
- Move the cursor to select 1:RegEQ. Press [ENTER]. This puts the regression equation into the Y= menu.

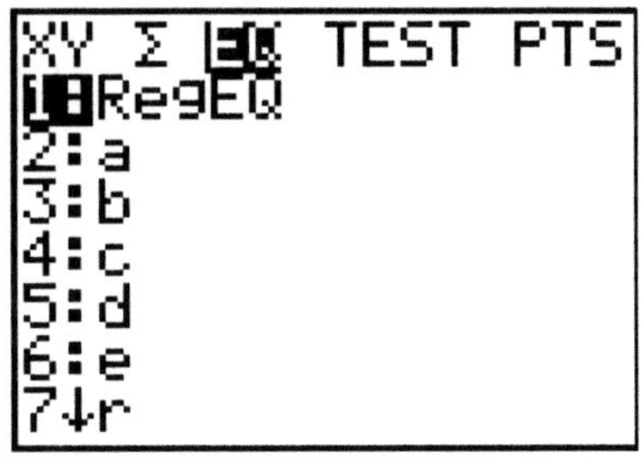

- Press the [GRAPH] key to see the regression equation and the scatter plot on the same screen.
- Press [TRACE] to trace the graphs.
- Use the up or down arrow to toggle between the scatter plot and the linear regression equation.

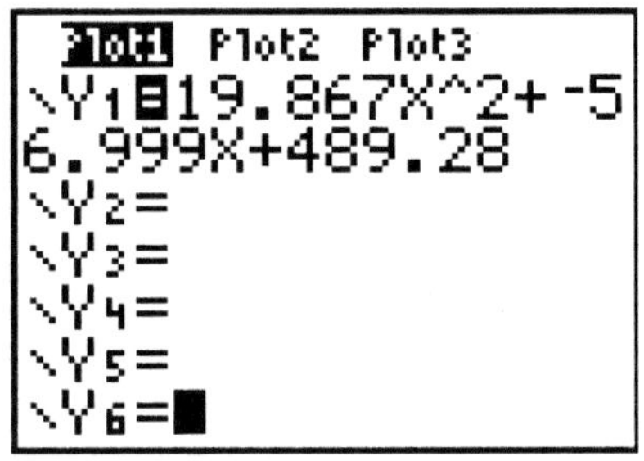

- To find the number of addicts on the quadratic model in 1991, first be sure that you are on Y1 and not Plot 1.
- Press the [TRACE] key. Enter 6 at the *X* prompt at the bottom of the screen. Press [ENTER] to see the *y*-coordinate when $x = 6$.
- Enter the value 15, to see the y-value 4104.37.

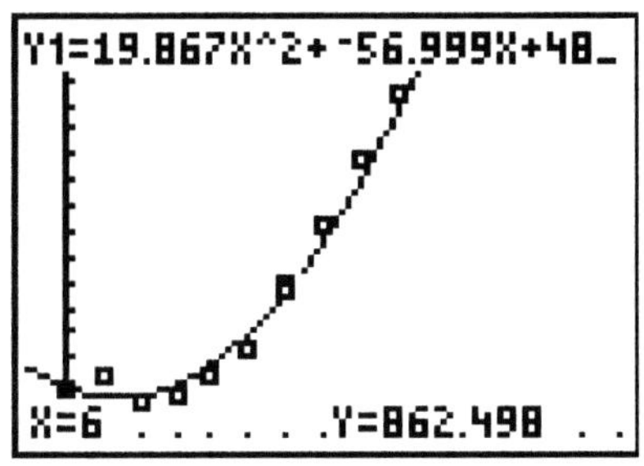

2.5 Example 1 on the TI-86

This example asks you to create a quadratic function that models the number of registered cocaine addicts in the United Kingdom for the years 1985-1995, using the number of years past 1985 as the input x.

Create the scatter plot - See 1.2 Example 5 for detailed instructions.

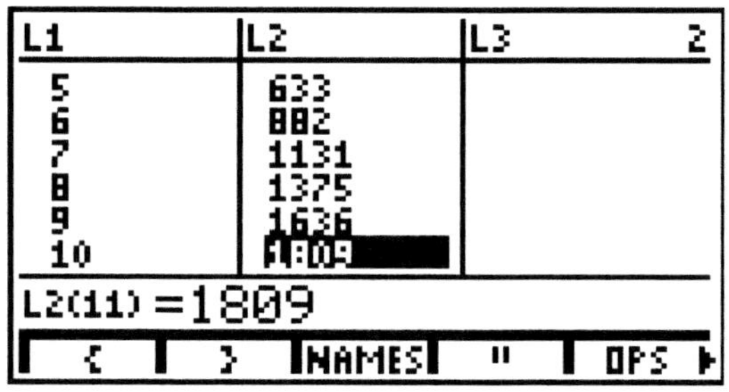

- Set the window to [-1, 15,1] by [100, 1900, 100]. Or use [F3] ZOOM then [MORE] [F5] ZDATA to set the window for the data.
- Use [2nd] [+] [F2] (EDIT) to see the STAT EDIT screen. Enter the x values into L1 and the number of addicts into L2. Reset the Mode to the word Float. The data will be displayed over two screens. The second screen is shown.

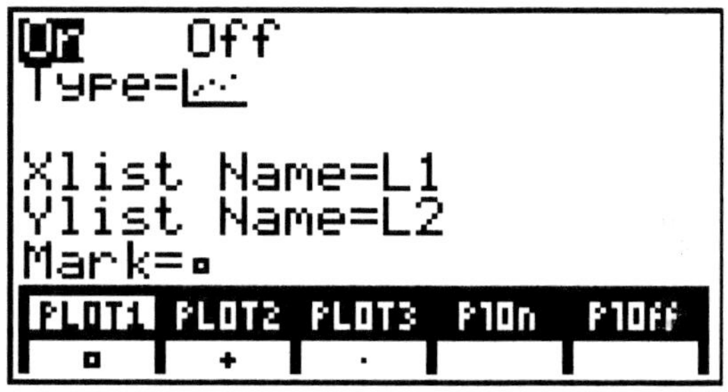

- Clear the lists if needed. Cursor to the name of the list to be cleared, press [CLEAR] [ENTER].
- Press [EXIT] [2nd] [+] to see the STAT menu again.
- Press [F3] (PLOT) then [F1] to format Plot 1 as shown. Be sure that the data is coming from L1 and L2.
- Press [EXIT] [EXIT] to return to the Home screen.

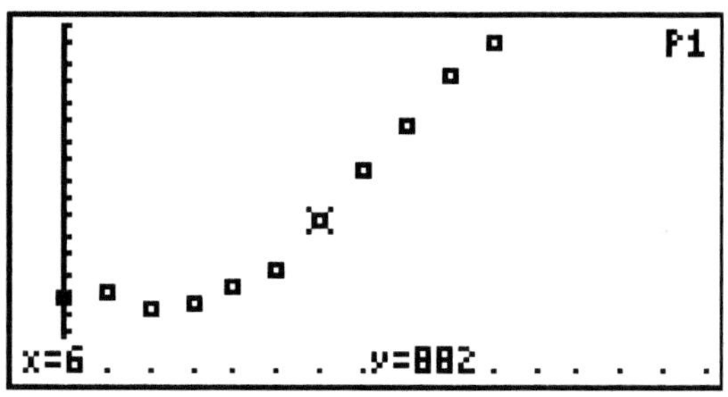

- Press the [GRAPH] key and turn off any y= functions. Press [EXIT] [F5] to see the scatter plot.
- Press [TRACE] to see the coordinates of the points. Use the right and left arrows to move from point to point.

Find the Quadratic Equation to Model the Data

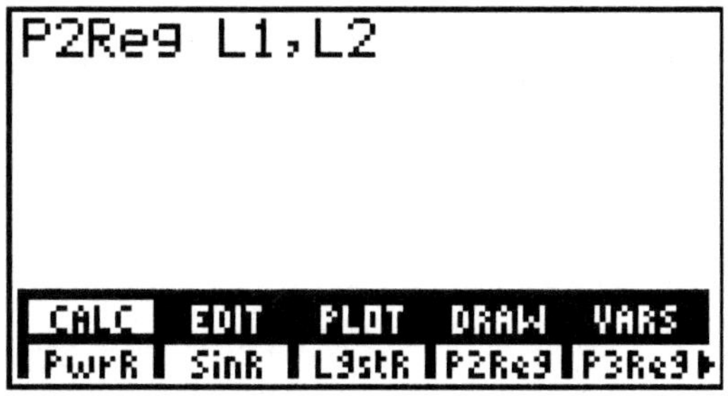

- Go back to the Home Screen. The data is already entered into L1 and L2, and Plot 1 is formatted for scatter plot.
- Go back to the STAT menu. Press [F1] to use CALC.
- Press [MORE] [F4] to select P2Reg. This puts P2Reg on the Home screen.
- You must now enter the names of the lists, L1 and L2 that contain the data. To do this, press [ALPHA] [7] [1] [,] [ALPHA] [7][2]. Press [ENTER] to see the values of a, b, and c as shown.

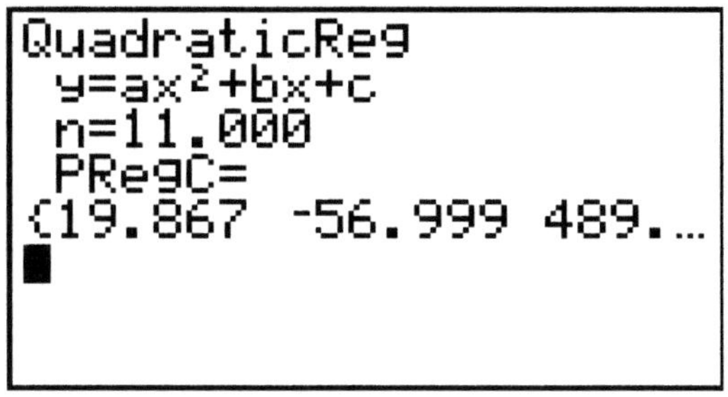

- Hence, the equation of the line through the data points is $y = 19.867x^2 - 56.999x + 489.280$.
- Use the [2nd][MORE] keys to access MODE. Change line 2 to Float 3 to round these values to 3 decimals as shown.

Graph the Quadratic Regression Equation on the Graph Screen

- Go to the graph menu. Turn off or clear all y = functions. Be sure that Plot 1 is on.
- Go back to the Home screen.
- Use [2nd][ENTER] to return to P2Reg L1, L2 screen. You may have to do this more than once. Immediately after L2 enter a comma and y1 using [,] [2nd] [ALPHA] [0] [1]. Note that you must use the lower case y.
- Press [ENTER] to see the regression on the Home screen.

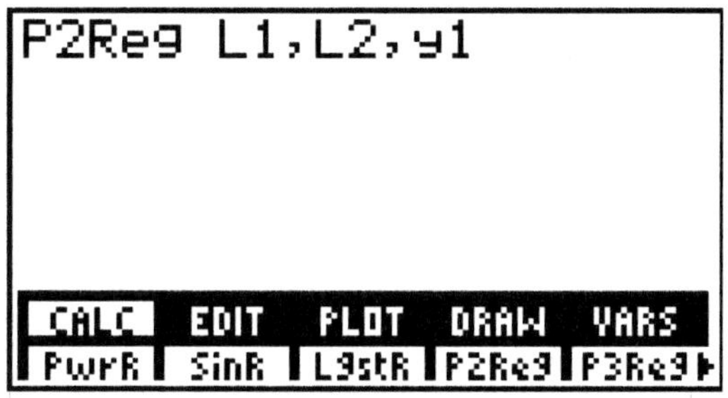

- Press the [GRAPH] key then [F1] to see the regression equation as y1 on the y= screen. Use the right arrow to scroll through the entire equation.
- Press [EXIT] [F5] to see the scatter plot and the linear regression equation graphed on the same screen. Press [TRACE] to trace the graphs.

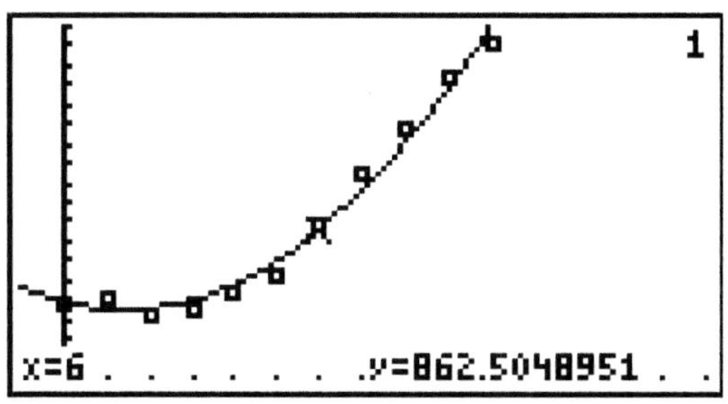

- Use the up or down arrow to toggle between the scatter plot and the linear regression equation.
- To find the number of addicts on the quadratic model in 1991, use [MORE][MORE][F1] to select EVAL. Enter 6 at the EVAL x = prompt at the bottom of the screen. Press [ENTER] to see the y-coordinate when $x = 6$. Repeat to find the y-value when x = 15.

2.5 Example 1 using EXCEL

This example asks you to create a quadratic function that models the number of registered cocaine addicts in the United Kingdom for the years 1985-1995, using the number of years past 1985 as the input x.

Create the scatter plot - See 1.2 Example 5 for detailed instructions.

- Enter labels and data for two columns in EXCEL, namely x and the Number of Addicts. You can use a third column for the actual year if you wish. Note that you must use x, the number of years since 1985, rather than the actual year to obtain the correct equation.
- Use the **Chart Wizard** to create the scatter plot.

Find the regression equation and put it on the scatter plot.

- Click anywhere on the scatter plot. This puts **Chart** on the menu bar.
- Go to **Chart** on the menu bar, select **Add Trendline**.
- Select Polynomial Order 2.
- Click on the Options tab and check Display equation on chart.
- Click **OK.**

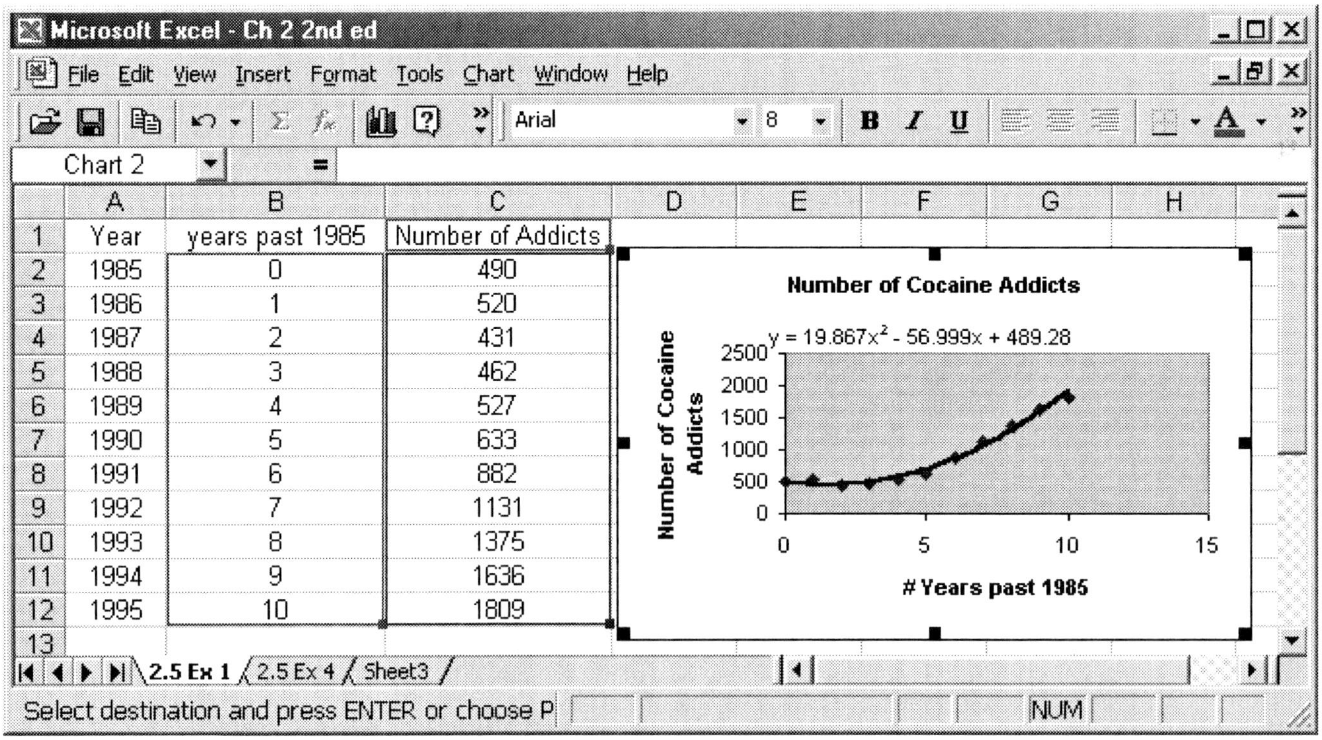

Evaluate a Formula

- First enter a label for the first two columns, say x in cell B19 and y in cell C19.
- Enter the value of x, say 6, into cell B20 and the equation as

  $=19.867 * B20 \wedge 2 - 56.999 * B20 + 489.280$ into cell C20.

  Note that you can click on cell B20 to put it in the formula.
- Press **ENTER.** You will see the value 862.498 in cell C20.
- To evaluate at other values of x, grab the right handle of the cell with the formula, C20, and drag it down as many rows as values you wish to evaluate. Change the value of x in the corresponding row and press **ENTER.**

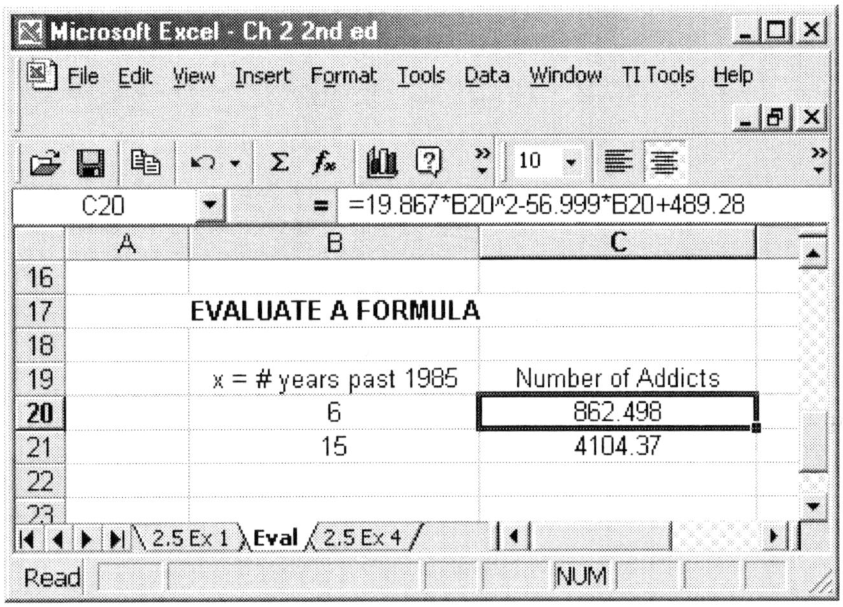

2.5 Example 4 on the TI-83, TI-83 PLUS, and TI-84 PLUS

This example asks you to find a power function that models the number of cohabiting households (in thousands) for selected years from 1960 to 1998. Let x represent the number of years from 1950, and y represent thousands of households.

The work in this example is the same as the work for 2.5 Example 1. The only difference is in that we model this data with a power function rather than a quadratic function.

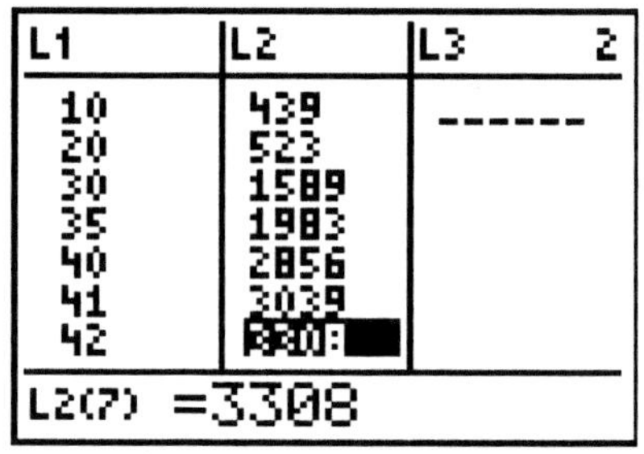

Create the scatter plot

- Enter the data into lists L1 and L2. Note that $x = 0$ is the year 1950, so the first data point has $x = 10$ for the year 1960. The list will span two screens. The first screen is shown.
- Examine the data to set the Window at [-10, 60, 10] by [-500, 4500, 500].
- Press GRAPH to see the scatter plot of the data.

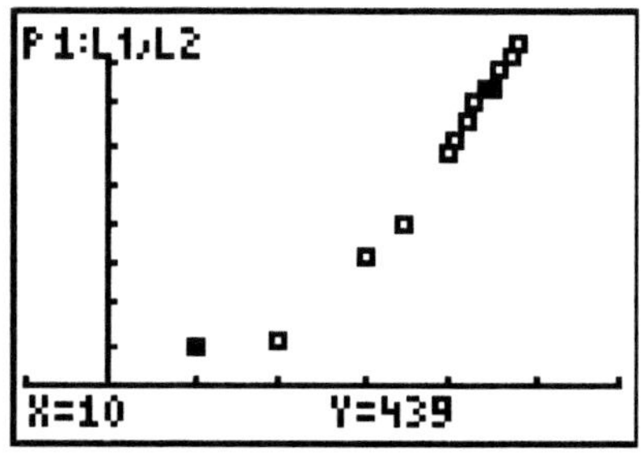

Find and graph the power function model for this data

- Go to the STAT CALC menu. Select A: PwrReg, the power function. Enter the names of the lists that contain the data.
- Press ENTER to see the a and b of the power function. The equation is $y = 6.999x^{1.634}$.
- Put the regression equation into the Y= menu. Place the cursor on Y1. Press VARS, select 5:Statistics EQ 1:RegEQ.
- Press ENTER to paste the equation as Y1.
- Graph both the scatter plot and the power function on the same screen.

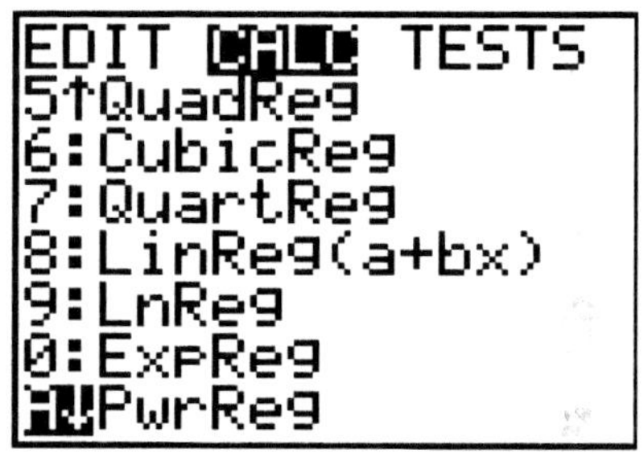

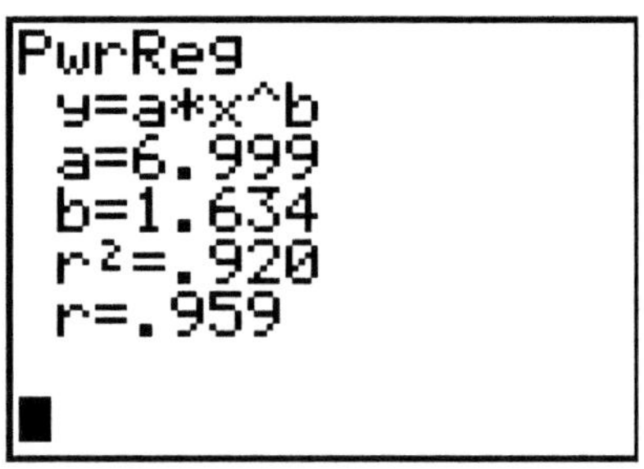

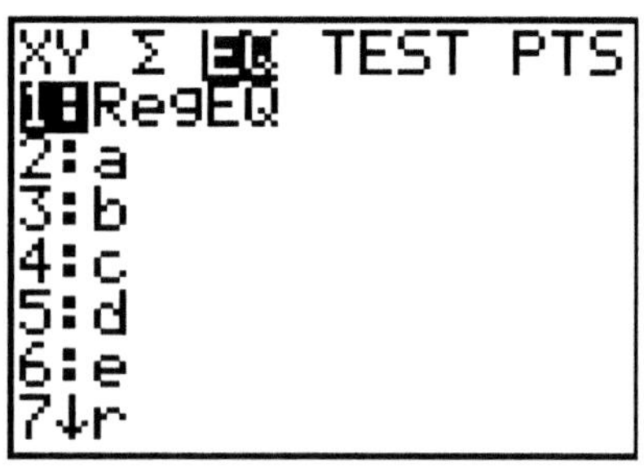

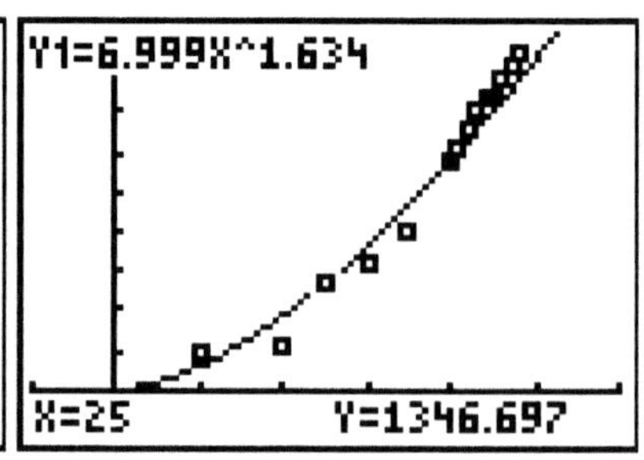

2.5 Example 4 on the TI-86

This example asks you to find a power function that models the number of cohabiting households (in thousands) for selected years from 1960 to 1998. Let x represent the number of years from 1950 and y represent thousands of households.

The work in this example is the same as the work for 2.5 Example 1. The only difference is in that we model this data with a power function rather than a quadratic function.

Create the scatter plot

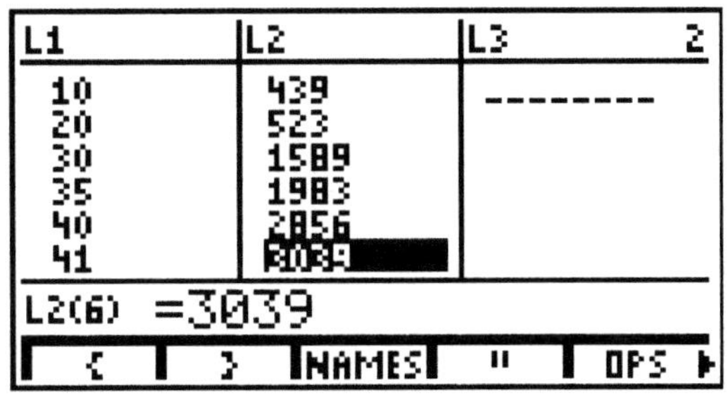

- Enter the data into lists L1 and L2. Note that $x = 0$ is the year 1950, so the first data point has $x = 10$ for the year 1960. The lists will span three screens.

  The first screen is shown.

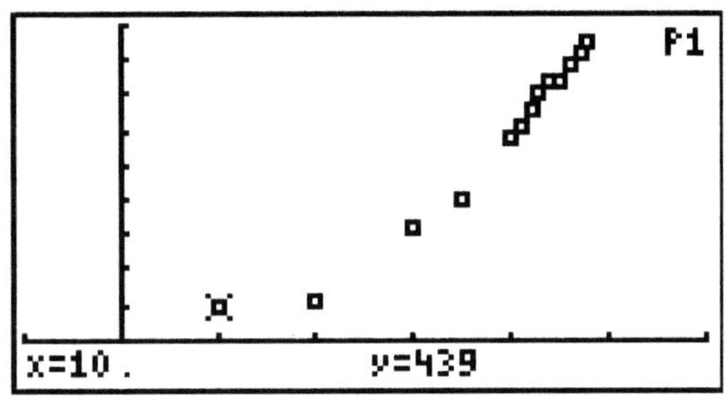

- Examine the data to set the Window at [-10, 60, 10] by [-500, 4500, 500].
- Graph the scatter plot of the data.

Find and graph the power function model for this data

- Go to the STAT CALC menu, press [MORE] once.
- Press [F1] to select PwrR, the power function.
- Enter the names of the lists that contain the data, and the location of a y= function. This will allow us to graph both the power function model and the scatter plot on the same screen.

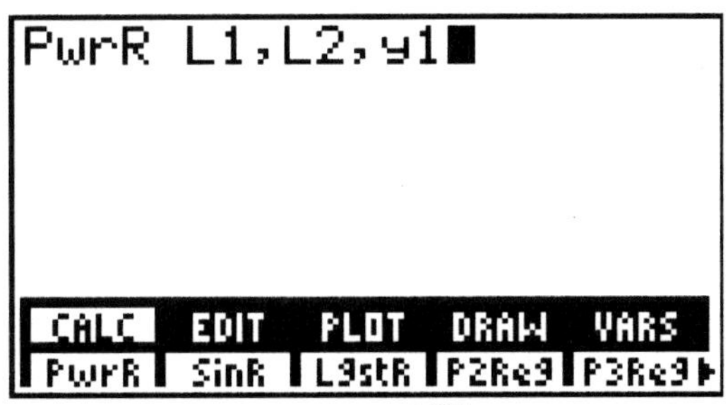

- Press [ENTER] to see the a and b of the power function. The equation is $y = 6.999x^{1.634}$.
- Graph both the scatter plot and the power function on the same screen.

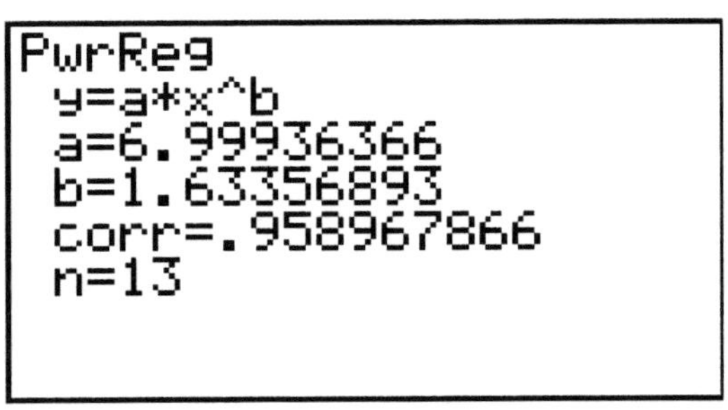

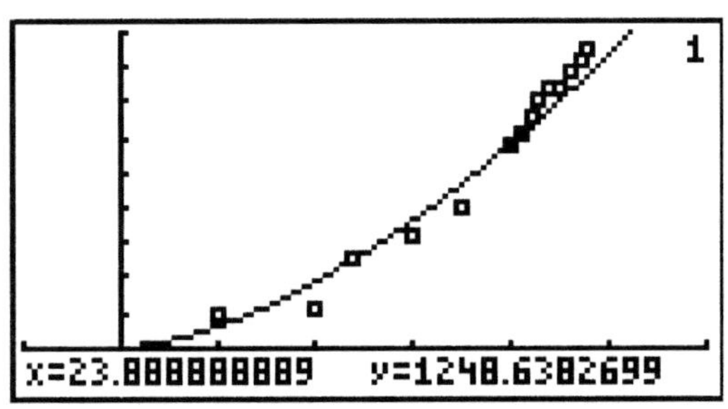

2.5 Example 4 using EXCEL

This example asks you to find a power function that models the number of cohabiting households (in thousands) for selected years from 1960 to 1998. Let x represent the number of years from 1950, and y represent thousands of households.

The work in this example is the same as the work for 2.5 Example 1. The only difference is in that we model this data with a power function rather than a quadratic function.

- Enter the data as shown.
- Create a scatter plot.
- Go to Chart Add Trendlines. Select Power.
- Click on the **Options** tab and select **Display equation.**
- The scatter plot and graph of the power function model are shown below.

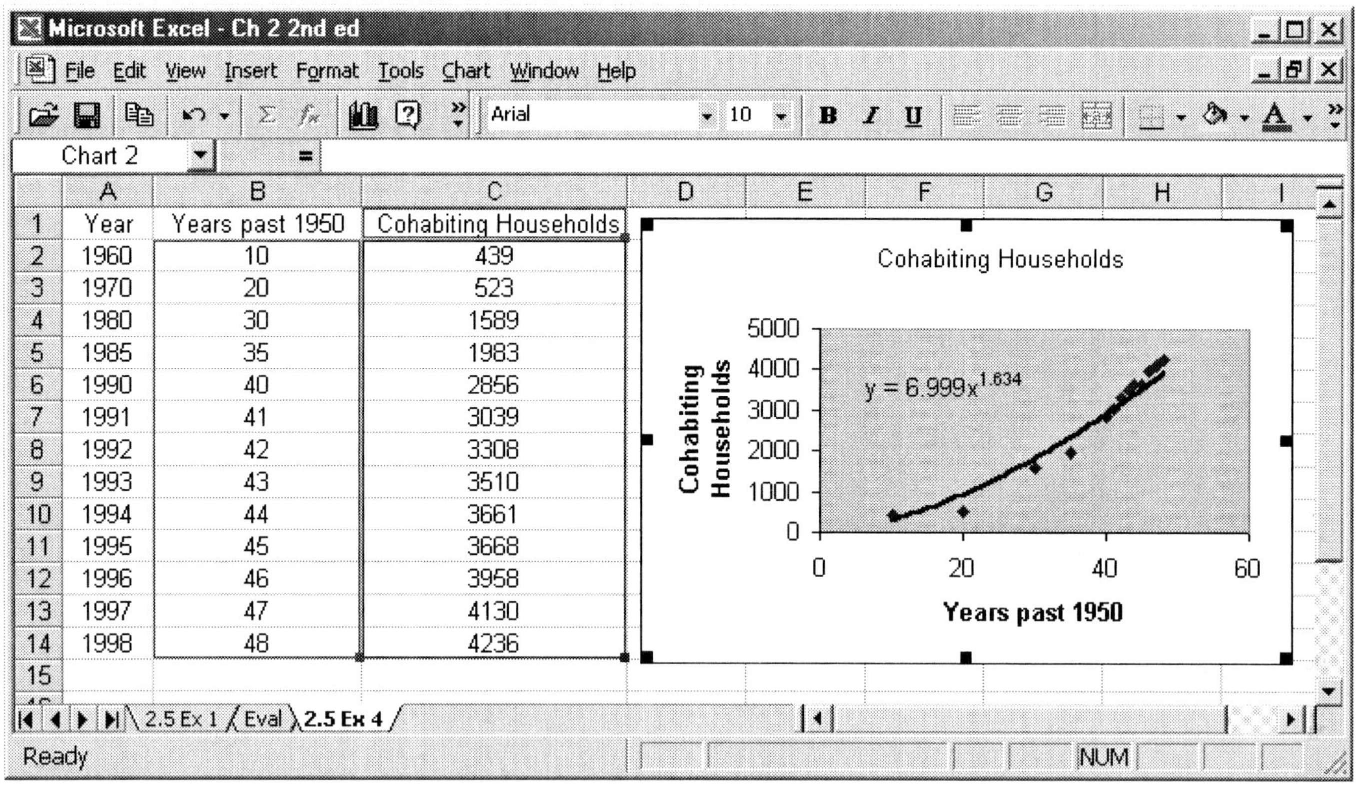

Year	Years past 1950	Cohabiting Households
1960	10	439
1970	20	523
1980	30	1589
1985	35	1983
1990	40	2856
1991	41	3039
1992	42	3308
1993	43	3510
1994	44	3661
1995	45	3668
1996	46	3958
1997	47	4130
1998	48	4236

2.6 Example 7 on the TI-83, TI-83 Plus, and TI-84 PLUS

This example shows how to combine functions by addition, division, and composition. Enter the individual functions in the Y= menu and then use the [VARS] key to use the names of the functions in the combined function.

If $f(x)=\sqrt{x-5}$ and $g(x)=2x^2-4$

Graph $(f+g)(x)$ in the window [-1, 10, 1] by [0, 200, 50].

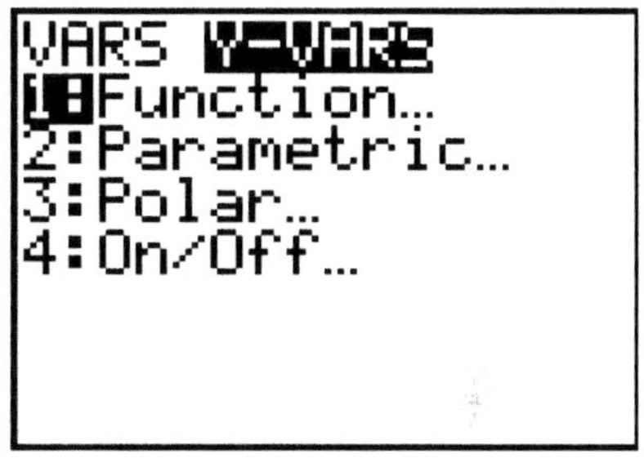

- Set the given window.
- Enter Y1 = $\sqrt{\ }(x-5)$ and Y2= $2x^2-4$.
- Place the cursor on Y3 and enter Y3 = Y1 + Y2.
- To do this, press the [VARS] key.
- Use the right arrow to go to the Y-VARS screen.
- Select 1:Function.
- Select Y1.
- Press the [+] key and repeat the process to select Y2.
- Turn off Y1 and Y2 to just see the graph of Y3.

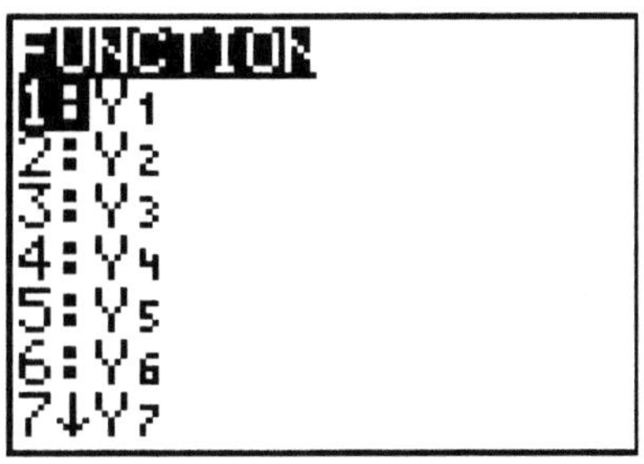

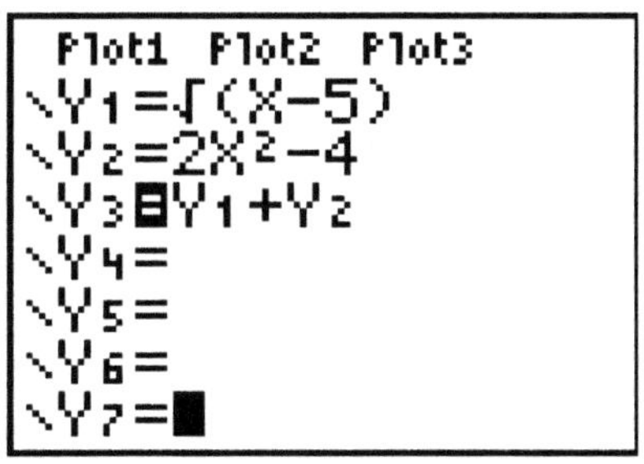

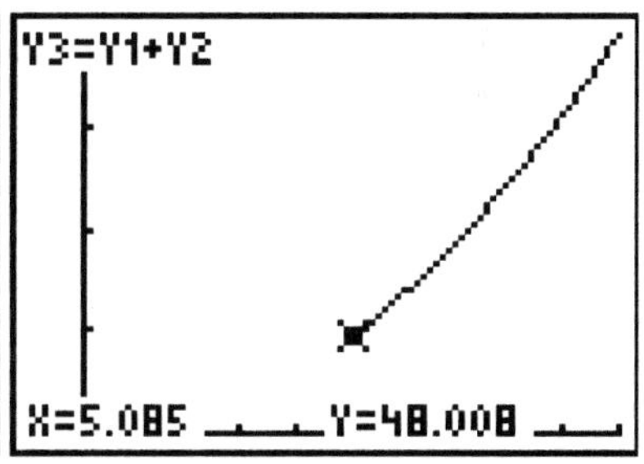

Graph $(f/g)(x)$ in the window [-1, 10, 1] by [0, 0.017, .001].

- Set the given window.
- Place the cursor on Y3 and enter Y3 = Y1 / Y2.
- Graph Y3.

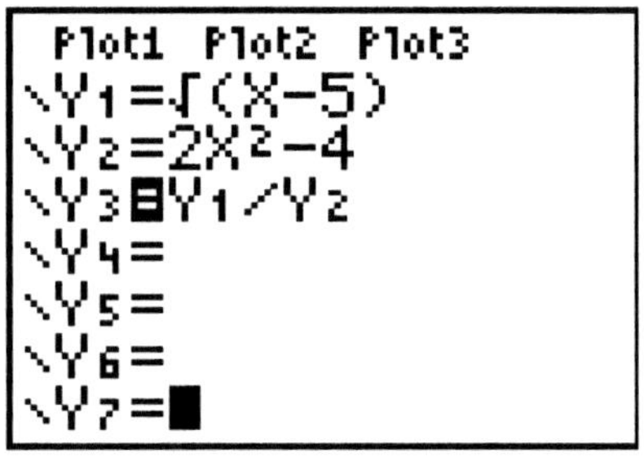

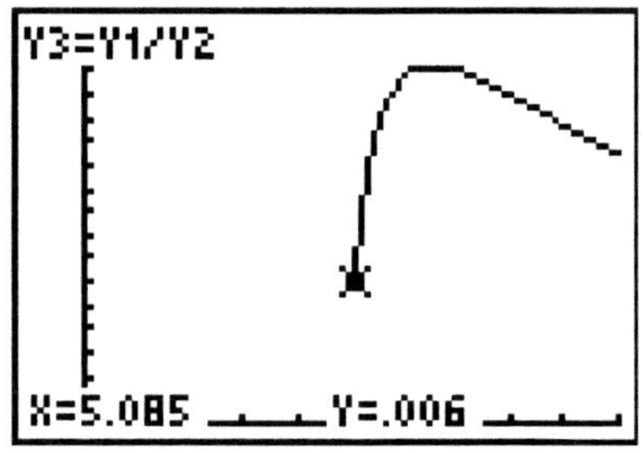

Compute $(f \circ g)(-3)$ and $(g \circ f)(9)$

- Press [2nd] [MODE] (QUIT) to return to the Home screen.
- Recall that $f(x) = \sqrt{\ }(x-5)$ is Y1 and $g(x) = 2x^2 - 4$ is Y2.
- To compute $(f \circ g)(-3)$, enter Y1(Y2(-3)).
- To enter Y1 and Y2 first press the [VARS] key.
- Use the right arrow to go to the Y-VARS screen.
- Select 1:Function, then select the appropriate function.
- Press [ENTER] to see the result, 3.

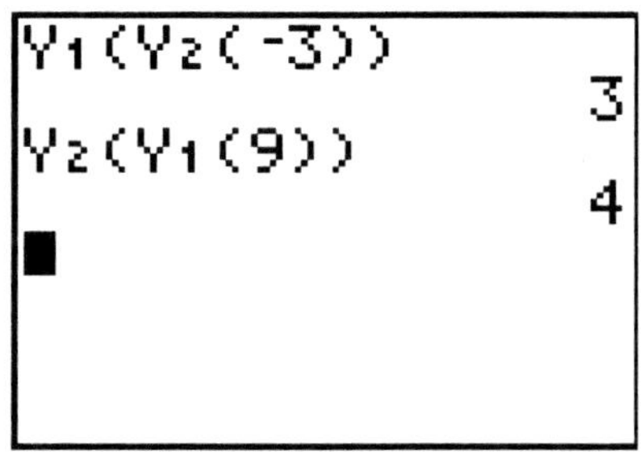

- To compute $(g \circ f)(9)$, enter Y2(Y1(9)).
- Press [ENTER] to see the result, 4.

2.6 Example 7 on the TI-86

This example shows how to combine functions by addition, division, and composition. Enter the individual functions in the Y= menu and then use the [VARS] key to use the names of the functions in the combined function.

If $f(x) = \sqrt{x-5}$ and $g(x) = 2x^2 - 4$

Graph $(f+g)(x)$ in the window [-1, 10, 1] by [0, 200, 50].

- Set the given window.
- Enter y1 = $\sqrt{\ }(x-5)$ and y2 = $2x^2 - 4$.
- Place the cursor on y3 and enter y3 = y1 + y2.
- You are in the y= menu. Press the [F2] key to place y on the screen.
- Turn off y1 and y2 to just see the graph of y3. To do this, place the cursor on the function you wish to turn off and press [F5] SELCT.
- Press [EXIT] [F5] GRAPH to see the graph of the sum of y1 + y2.

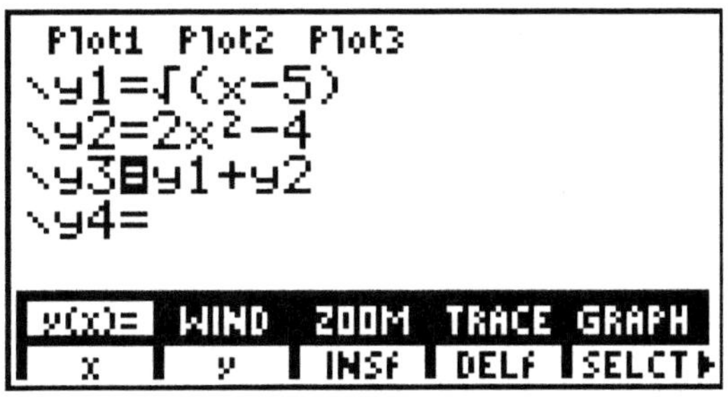

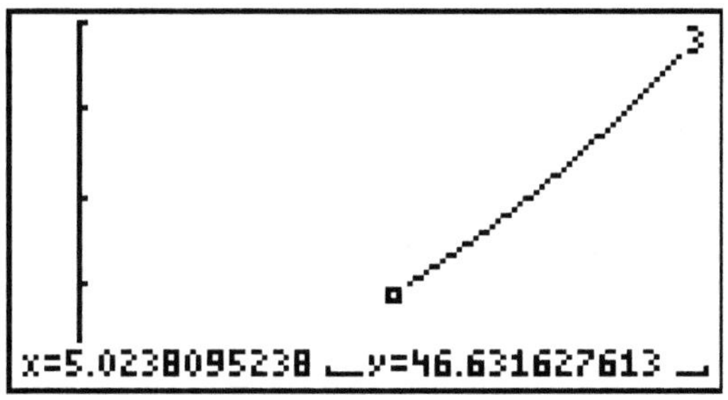

Graph $(f/g)(x)$ in the window [-1, 10, 1] by [0, 0.017, .001].

- Set the given window.
- Place the cursor on y3 and enter y3 = y1 / y2.
- Graph y3.

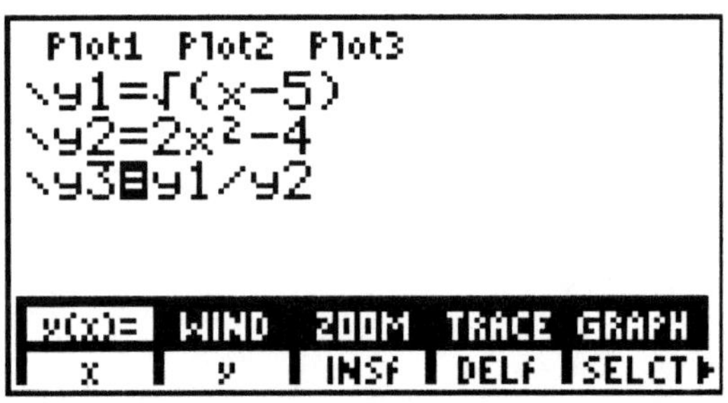

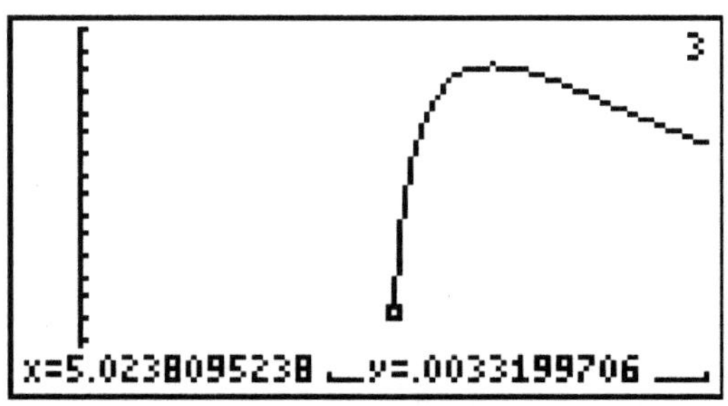

Compute $(f \circ g)(-3)$ and $(g \circ f)(9)$

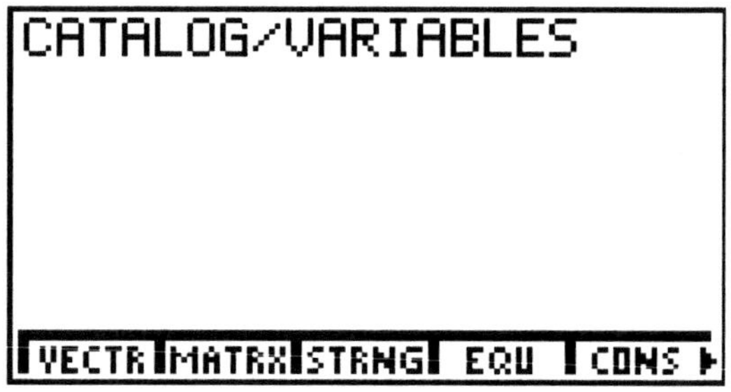

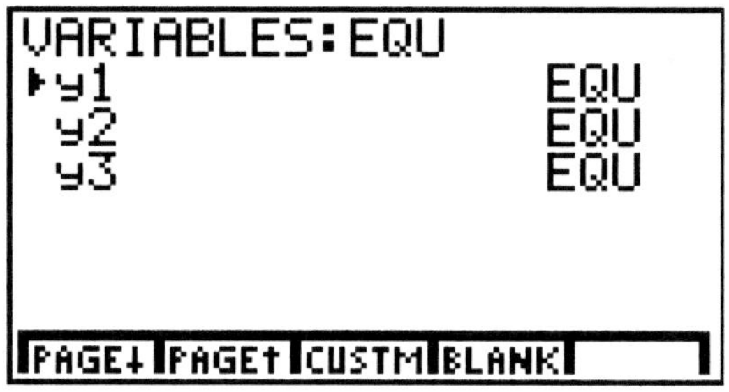

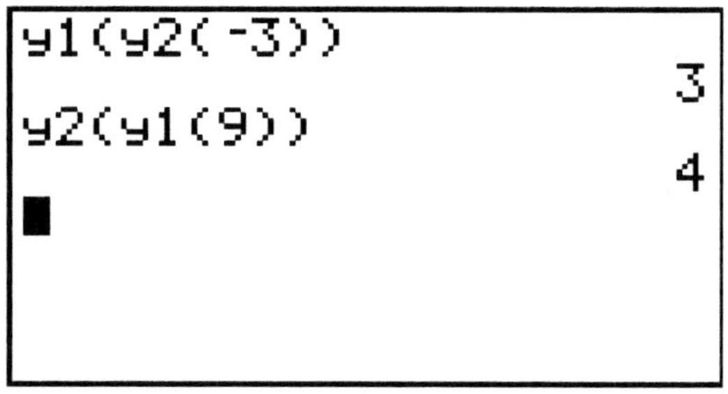

- Press [EXIT] [EXIT] to return to the Home screen.
- Recall that $f(x) = \sqrt{\ }(x-5)$ is y1 and $g(x) = 2x^2 - 4$ is y2.
- To compute $(f \circ g)(-3)$, enter y1(y2(-3)).
- To enter y1 and y2 first press [2nd] [CUSTOM] [MORE] [F4] to select EQU.
- Place the cursor on the function you want. Press [ENTER]
- Repeat to enter the rest of the expression.
- Press [ENTER] to see the result, 3.
- To compute $(g \circ f)(9)$, enter y2(y1(9)).
- Press [ENTER] to see the result, 4.

Put y1 and y2 into CUSTOM

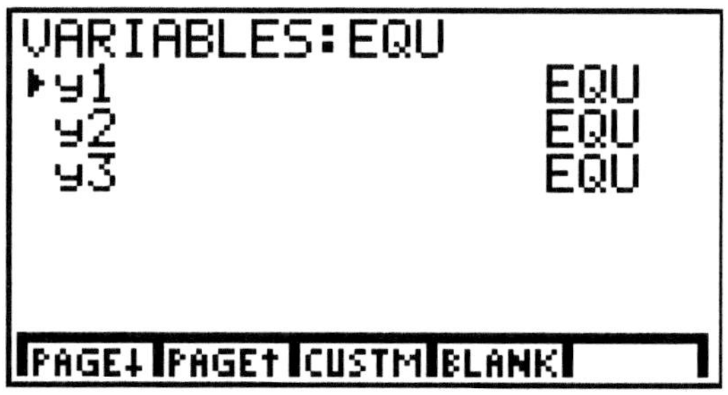

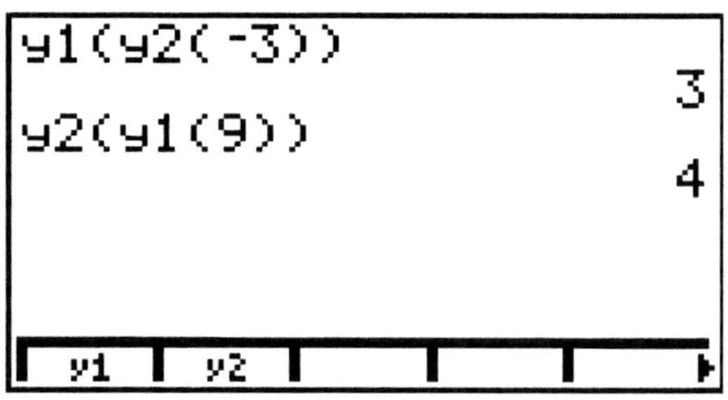

- You can put y1 and y2 into the CUSTOM menu for easy access.
- Press [2nd] [CUSTOM] [MORE] [F4] to select EQU.
- Place the cursor on y1.
- Press [F3] to see the CUSTOM menu.
- Press [F1] to place y1 in the first box in the CUSTOM menu.
- Repeat to place y2 in [F2].
- Press [EXIT] to return to the Home screen.
- To place y1 on the Home screen, press the [CUSTOM] key and [F1].
- To place y2 on the Home screen, press the [CUSTOM] key and [F2].

CHAPTER 3

EXPONENTIAL AND LOGARITHMIC FUNCTIONS

3.4 Example 2 on the TI-83, TI-83 Plus and TI-84 PLUS

This example asks you to create an exponential function that models the percent of U.S. males 18 years of age and older that smoked in selected years from 1965 to 1995. The number of years after 1960 is x, so the first data point of 1965 is $x = 5$.

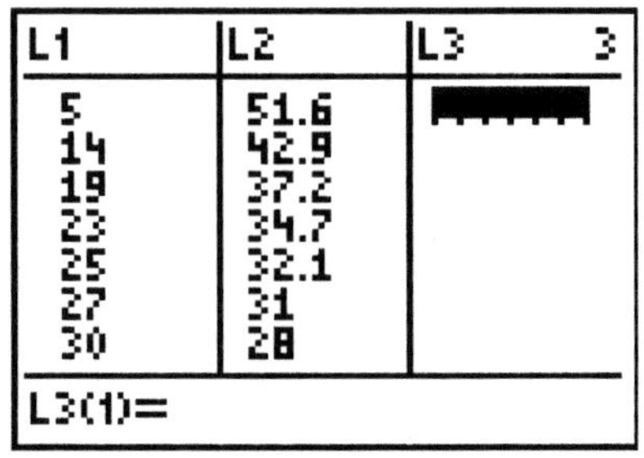

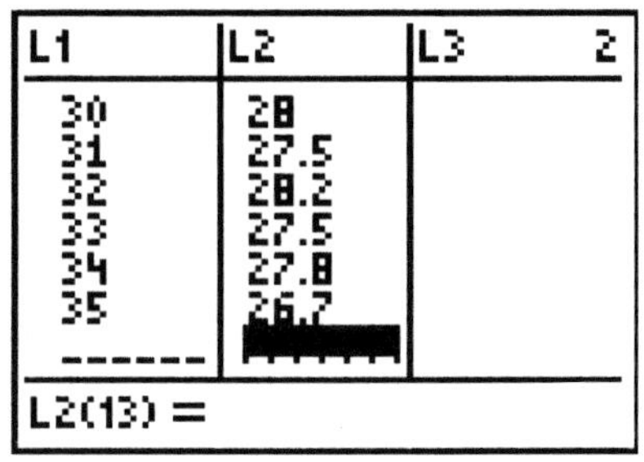

Create the scatter plot.

- First convert each year in the data to an x value.
- Set the window to [0, 40, 5] by [0, 70, 10]. Or use [ZOOM] 9: ZoomStat to set the window for the data.
- Use [STAT] EDIT 1: Edit to enter the x values into L1 and the percentage of smokers into L2. Set the Mode to the word Float.
- Clear the lists if needed. Cursor to the name of the list to be cleared, press [CLEAR] [ENTER].
- Press [2nd] [Y=] to turn on and format Plot 1. Be sure that the data is coming from L1 and L2.
- Press the [GRAPH] key and then [TRACE] to see the scatter plot. Use the right and left arrows to move from point to point.

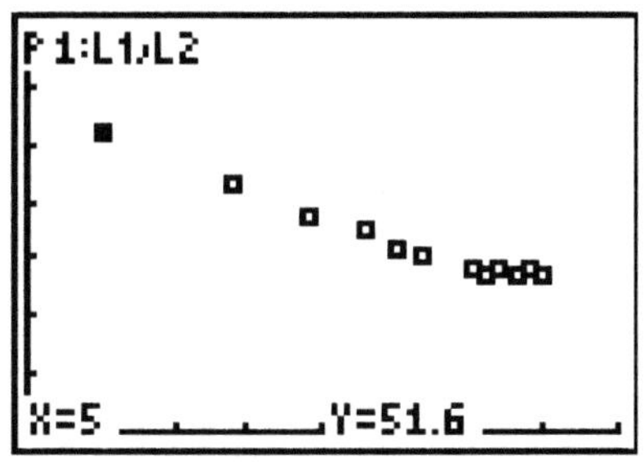

Find the Exponential Equation to Model the Data.

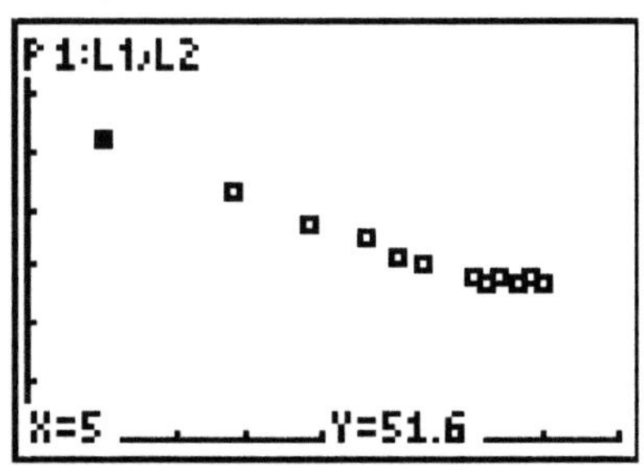

- Return to the Home Screen.
- The data is already entered into L1 and L2, and Plot 1 is formatted for scatter plot.
- Reset the Mode to Float 3. Otherwise, you can round by hand.
- Next, press the [STAT] key and use the right arrow to go to CALC.
- Select 0: ExpReg. This puts ExpReg on the Home screen.

- Press [ENTER] to see the values of a and b as shown.
- Hence, the equation of the exponential model through the data points is $y = 57.66(0.978^x)$.
- Turn DiagnosticOn if you do not see r^2.

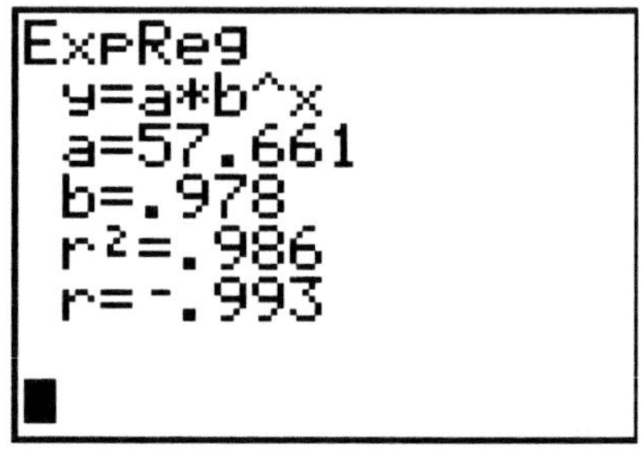

Graph the Exponential Regression Equation on the Graph Screen

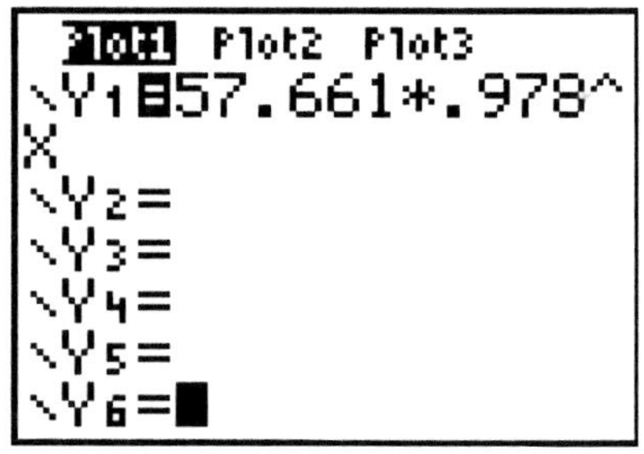

- Turn off all Y= functions. Be sure that Plot 1 is on.
- Move the cursor to the Y= line you wish to use.
- Press the [VARS] key. Select 5:Statistics.
- Move the right arrow twice to see the EQ menu.
- Move the cursor to select 1:RegEQ. Press [ENTER]. This puts the regression equation into the Y= menu.
- Press the [GRAPH] key to see the regression equation and the scatter plot on the same screen.
- Press [TRACE] to trace the graphs.
- Use the up or down arrow to toggle between the scatter plot and the exponential regression equation.

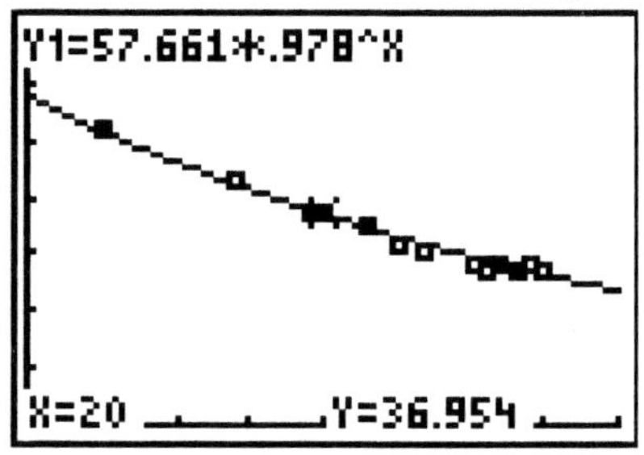

- To find the percent of smokers in 2000, use [2nd] CALC 1:value. Enter 40 at the *X* prompt at the bottom of the screen. Press [ENTER] to see the *y*-coordinate when $x = 40$.
- Notice that the value of *y* is different from 23.3 as stated in the text. This is because we set the Float to 3 and rounded the values of a and b before we pasted it into Y1. The calculator then used only the rounded values. To get the actual value of 23.3 you must reset the mode to the Float and go back and redo the regression equation.

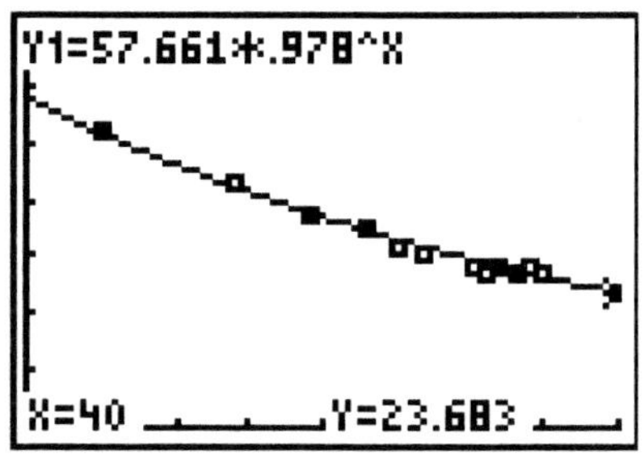

3.4 Example 2 on the TI-86

This example asks you to create an exponential function that models the percent of U.S. males 18 years of age and older that smoked in selected years from 1965 to 1995. The number of years after 1960 is x, so the first data point of 1965 is $x = 5$.

Create the scatter plot

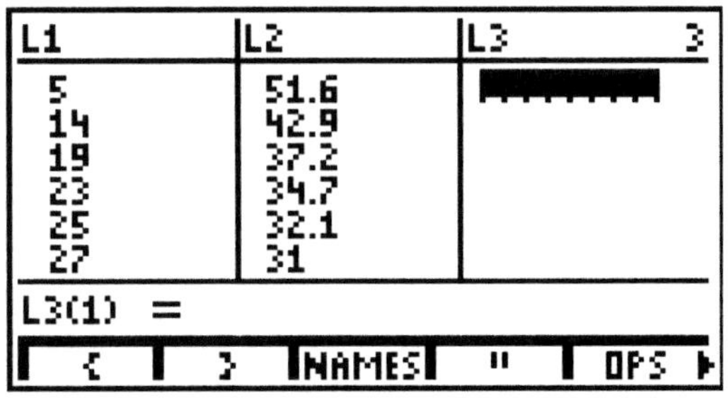

- First convert each year in the data to an x value.
- Set the window to [0, 40, 5] by [0, 70, 10]. Or use [F3] ZOOM then [MORE] [F5] ZDATA to set the window for the data.
- Use [2nd] [+] [F2] to see the STAT EDIT screen. Enter the x values into L1 and the percents into L2. The first screen is shown.
- Clear the lists if needed. Cursor to the name of the list to be cleared, press [CLEAR] [ENTER].

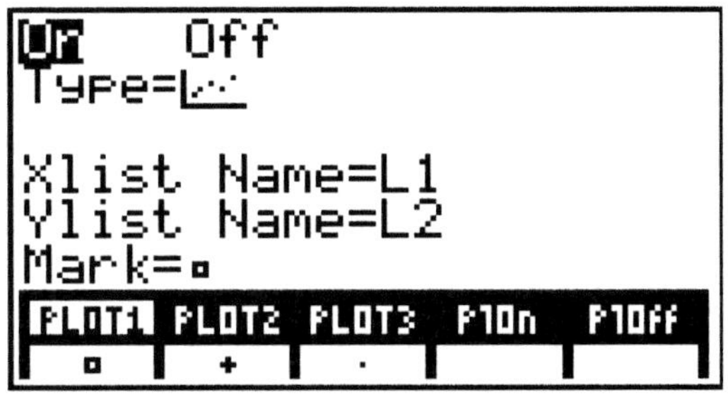

- Press [EXIT] [2nd] [+] to see the STAT menu.
- Press [F3] then [F1] to format Plot 1 as shown. Be sure that the data is coming from L1 and L2.
- Press [EXIT] [EXIT] to return to the Home screen.

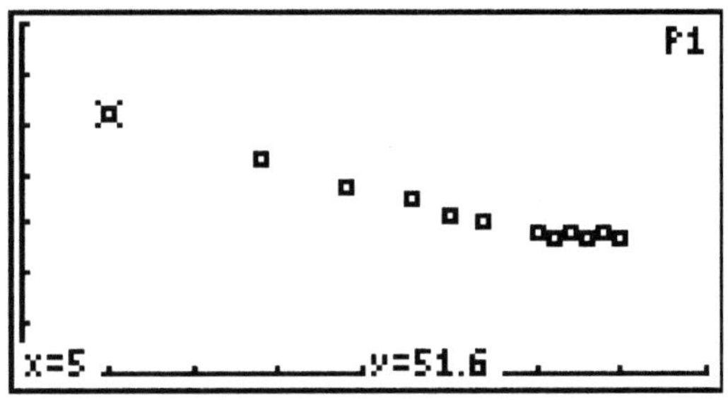

- Press the [GRAPH] key and turn off any y= functions. Press [F5] to see the scatter plot.
- Press [TRACE] to see the coordinates of the points. Use the right and left arrows to move from point to point.

Find the Exponential Equation to Model the Data.

- The data is already entered into L1 and L2, and Plot 1 is formatted for scatter plot.
- Return to the Home Screen.
- The data is already entered into L1 and L2, and Plot 1 is formatted for scatter plot.

- Reset the Mode to Float 3. Otherwise, you can round by hand.
- Go back to the STAT menu. Press [F1] to use CALC.
- Press [F5] to select ExpR. This puts ExpR on the Home screen.

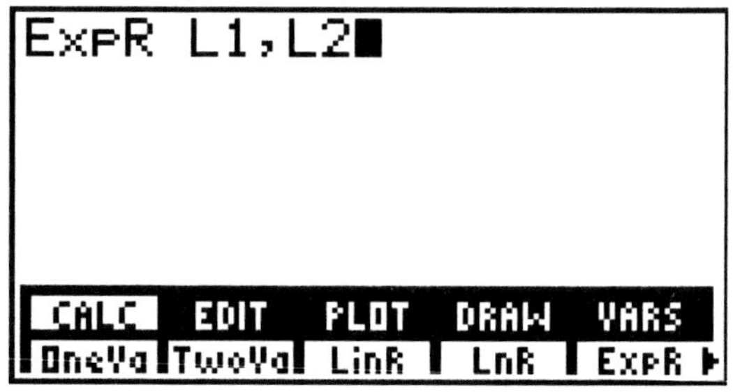

- You must now enter the names of the lists, L1 and L2 that contain the data. To do this, press [ALPHA] [7] [1][,] [ALPHA] [7][2].
- Press [ENTER] to see the values of a, b as shown.
- Hence, the equation of the line through the data points is $y = 57.66(0.978^x)$.

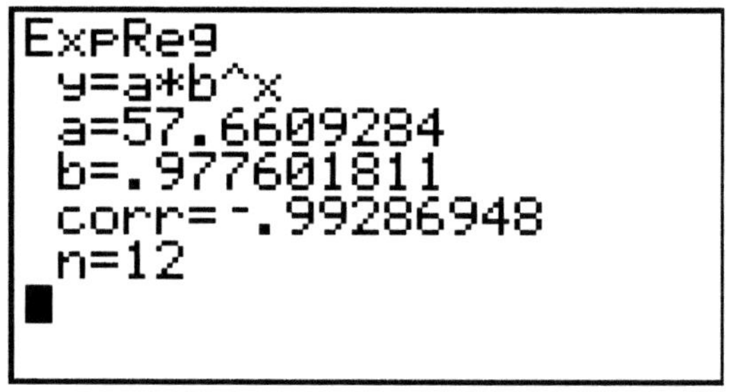

Graph the Exponential Regression Equation on the Graph Screen

- Go to the graph menu. Turn off or clear all y= functions. Be sure that Plot 1 is on.
- Go back to the Home screen.
- Use [2nd] [ENTER] to return to the ExpR L1, L2 screen. You may have to do this more than once. Immediately after L2 enter a comma and y1 using [CUSTOM] or [,] [2nd] [ALPHA] [0] [1]. Note that you must use the lower case *y*.

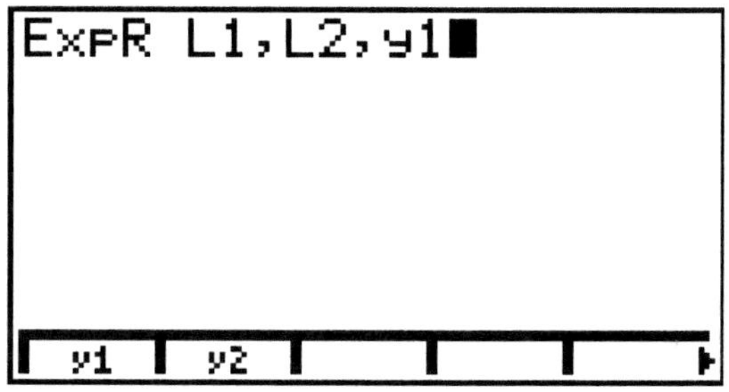

- Press [ENTER] to see the regression on the Home screen.
- Press the [GRAPH] key then [F1] to see the regression equation as y1 on the y= screen. Use the right arrow to scroll through the entire equation.
- Press [EXIT] [F5] to see the scatter plot and the exponential regression equation on the same screen.

- Press [TRACE] to trace the graphs.
- Use the up or down arrow to toggle between the scatter plot and the linear regression equation.
- To find the percent of smokers in the year 2000, use [MORE][MORE][F1] to select EVAL. Enter 40 at the EVAL x = prompt at the bottom of the screen. Press [ENTER] to see the y-coordinate when $x = 40$.

3.4 Example 2 using EXCEL

This example asks you to create an exponential function that models the percent of U.S. males 18 years of age and older that smoked in selected years from 1965 to 1995. The number of years after 1960 is x, so the first data point of 1965 is $x = 5$.

Create the scatter plot

- Enter labels and data for two columns in EXCEL, namely x and the percent of smokers. You can use a third column for the actual year if you wish. Note that you must use x, the number of years since 1960, rather than the actual year to obtain the correct equation.
- Use the **Chart Wizard** to create the scatter plot.

Find the regression equation and put it on the scatter plot.

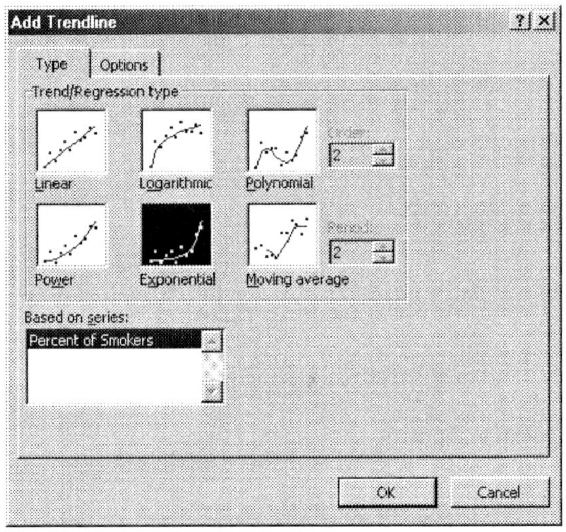

- Click anywhere on the scatter plot. This puts **Chart** on the menu bar.
- Go to **Chart** on the menu bar, select **Add Trendline**.
- Select Exponential.
- Click on the Options tab and check Display equation on chart.
- Click **OK.**

Evaluate a Formula

- First enter a label for the first two columns, say x in cell B16 and y in cell C16.
- Enter the value of x, say 40 into cell B17 and the equation as $= 57.66 * e \wedge (-0.0227 * B20)$ into cell C17. Note that you can click on cell B17 to put it in the formula.

- To enter the exponential function you must use the function listing of EXCEL, *fn*.
- On the left of the function listing, select **Math & Trig.**
- On the right scroll down until you find Exp. This is the exponential function with base *e*.
- Click **OK.**
- A dialog box appears. Fill it in as shown.

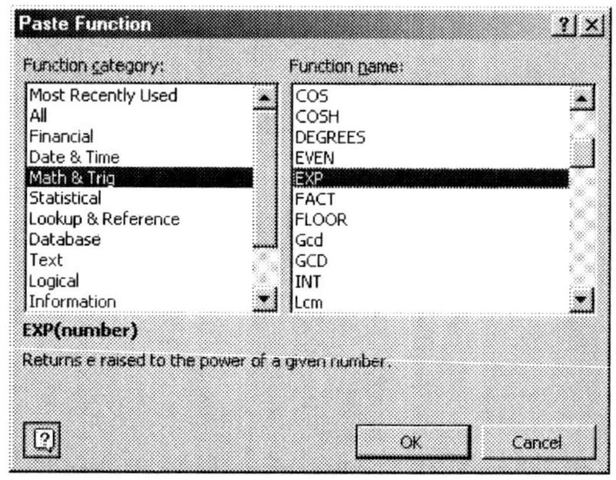

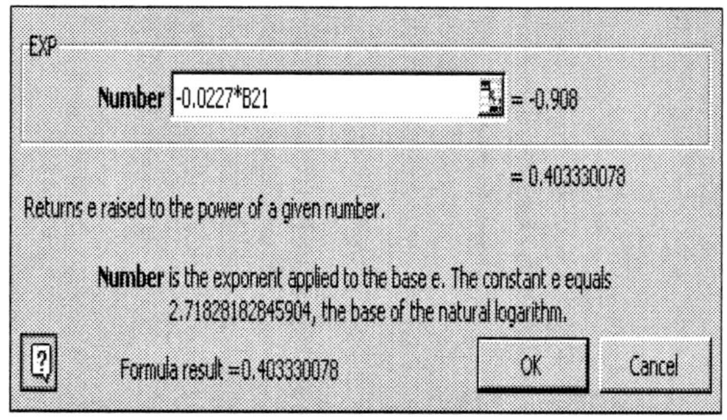

- Click **OK.**
- You will see the value 23.3 in cell C17 as shown on the EXCEL sheet. below
- To evaluate at other values of *x*, change the value of *x* and press **ENTER.**

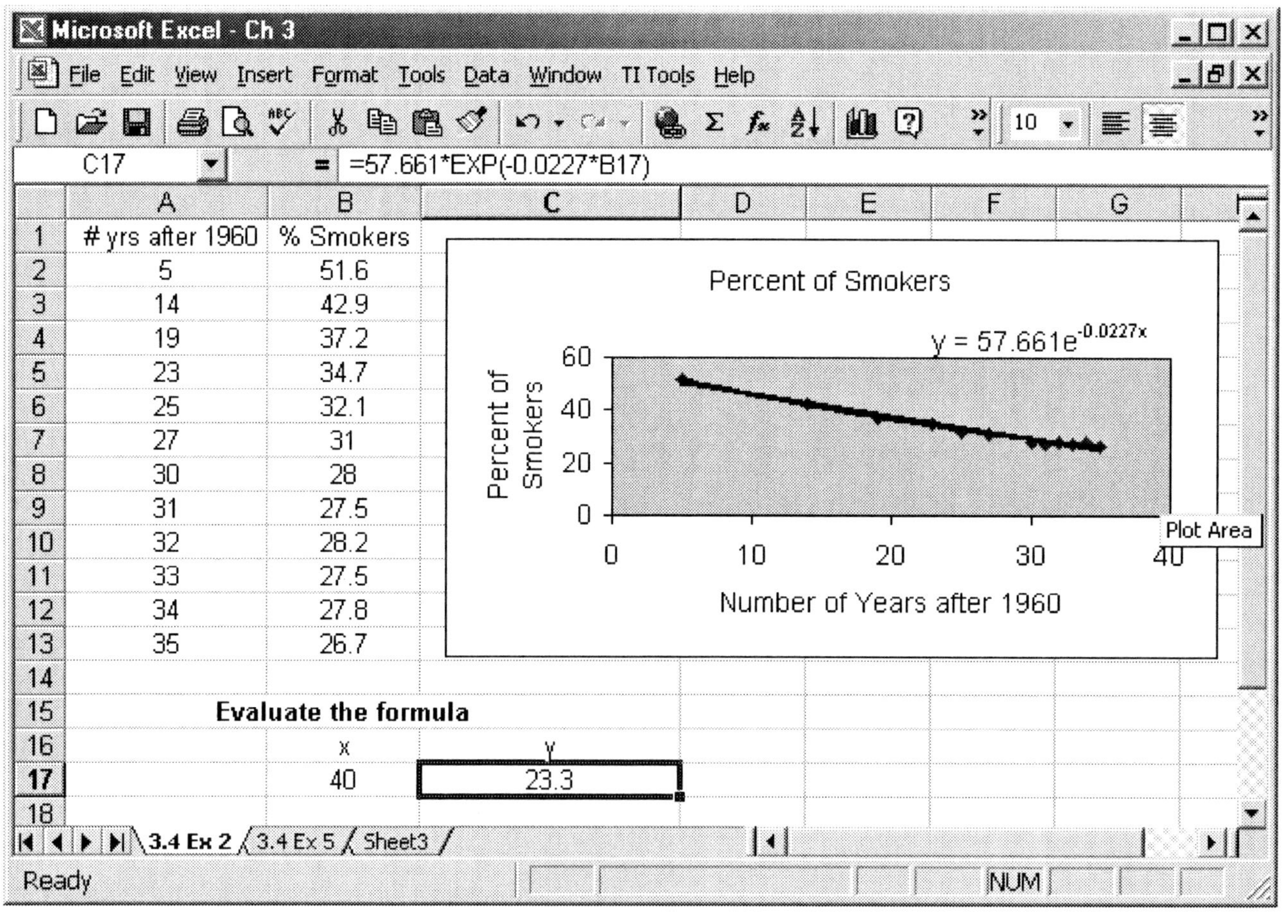

3.4 Example 5 on the TI-83 and TI-83 PLUS

This Example asks you to find a logarithmic function that models the percent of moms who returned to the work force within one year after having a child for the years from 1976 to 1998. Let x represent the number of years from 1970 and y represent percent of moms.

Create the scatter plot

- Enter the data into lists L1 and L2. Note that $x = 0$ is the year 1970, so the first data point has $x = 6$ for the year 1976.

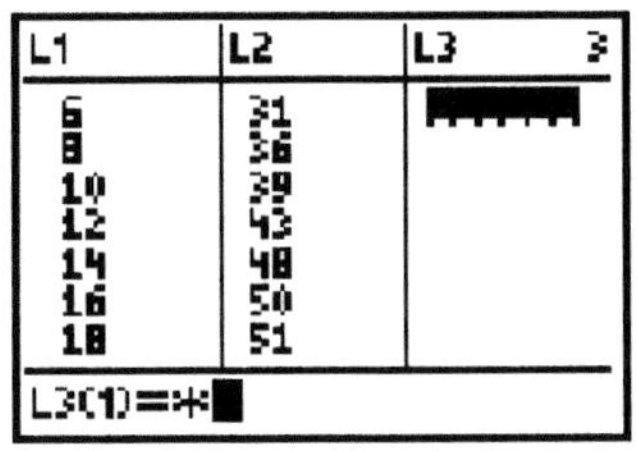

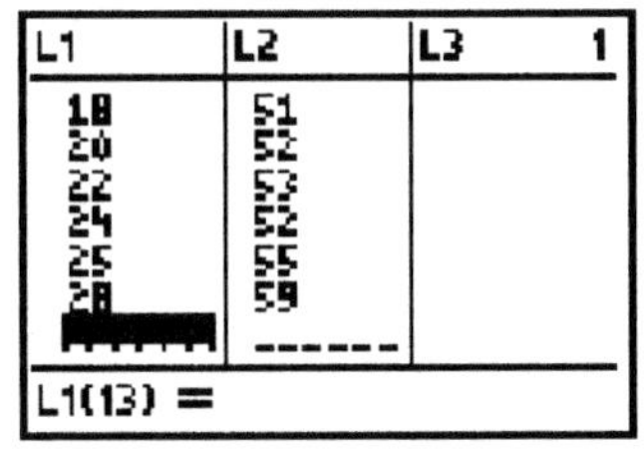

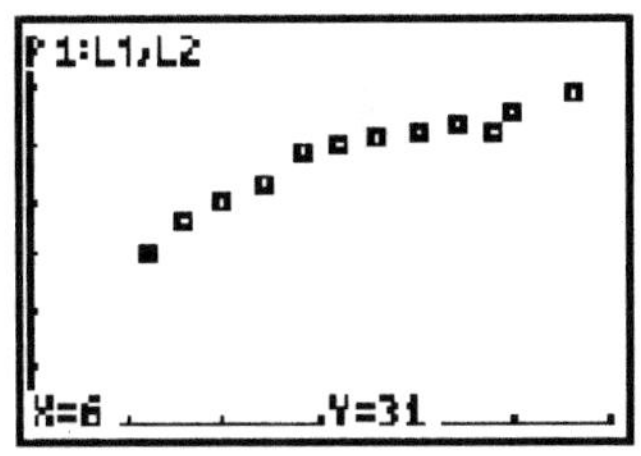

- Examine the data to set the Window at [0,30] by [0,70].
- Set Xscl = 5 and Yscl=10.
- Create the scatter plot of the data.

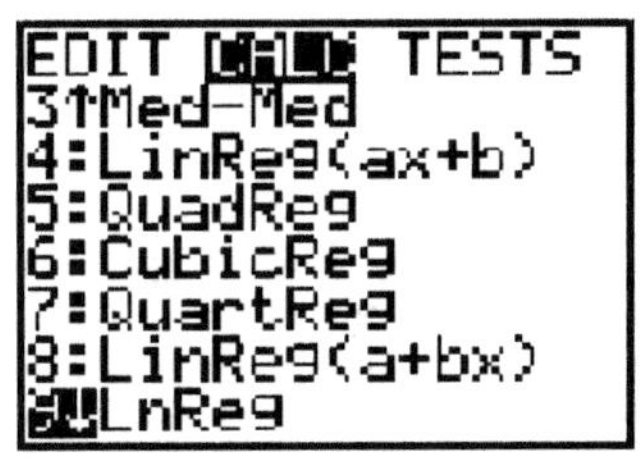

Find and graph the logarithmic function model for this data

- Go to the STAT CALC menu.
- Select 9:LnReg, the natural log function.
- Enter the names of the lists that contain the data.
- Press [ENTER] to see the a and b of the logarithmic model. The equation is $y = 0.815 + 17.062 \ln x$.

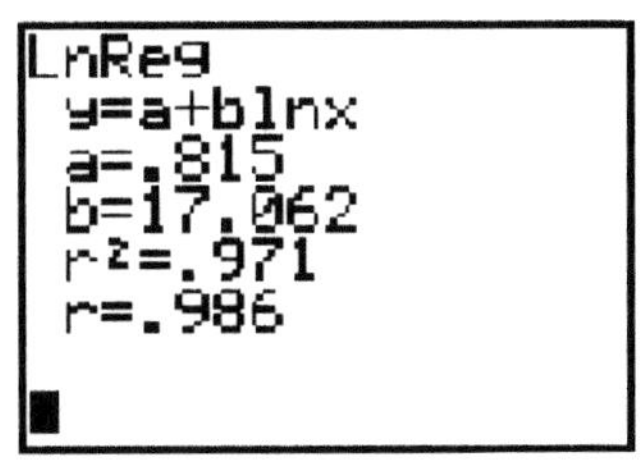

- Put the regression equation into the Y= menu. Place the cursor on Y1. Press [VARS], select 5:Statistics EQ 1:RegEQ.
- Press [ENTER] to paste the equation as Y1.
- Graph both the scatter plot and the logarithmic function on the same screen.

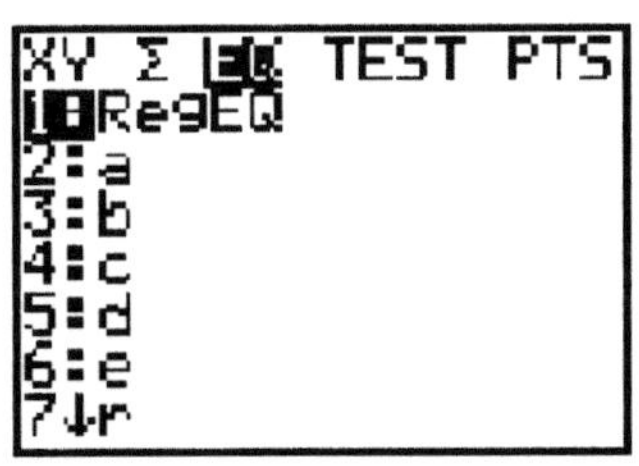

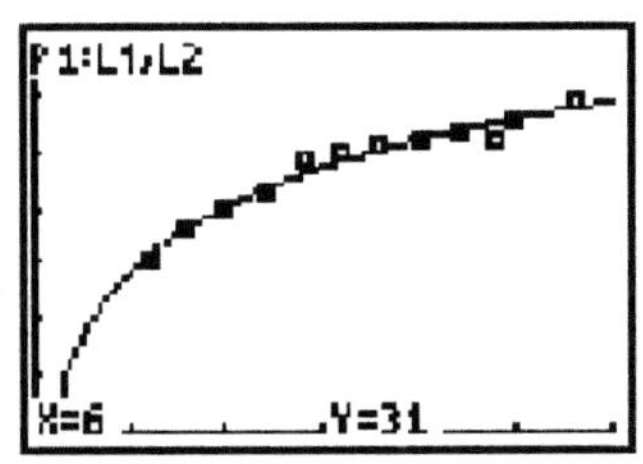

3.4 Example 5 on the TI-86

Example 5 asks you to find a logarithmic function that models the percent of moms who returned to the work force within one year after having a child (for the years from 1976 to 1998. Let x represent the number of years from 1970 and y represent percent of moms.

Create the scatter plot

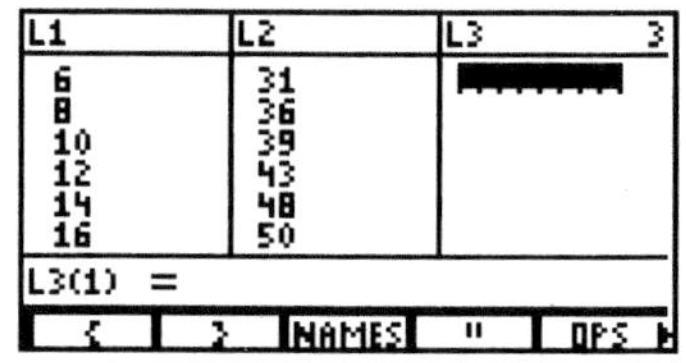

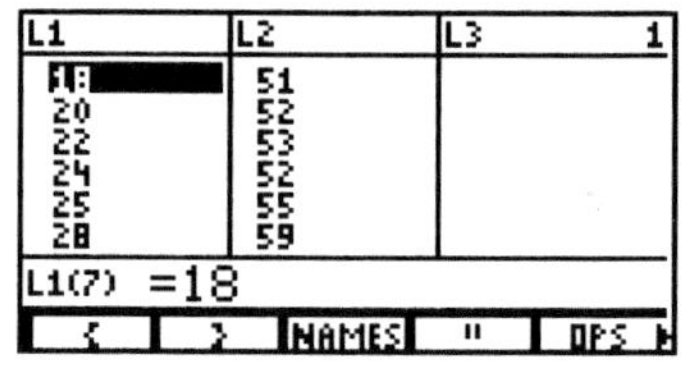

- Enter the data into lists L1 and L2. Note that $x = 0$ is the year 1970, so the first data point has $x = 6$ for the year 1976.
- Examine the data to set the Window at [0,30] by [0,70]. Set xScl = 5 and yScl=10.
- Create the scatter plot of the data.

Find and graph the logarithmic function model for this data

- Go to the STAT CALC menu.
- Press [MORE] once then [F4] to select LnR, the logarithmic function.
- Enter the names of the lists that contain the data, and the location of a y = function. This will allow us to graph both the logarithmic function model and the scatter plot on the same screen.

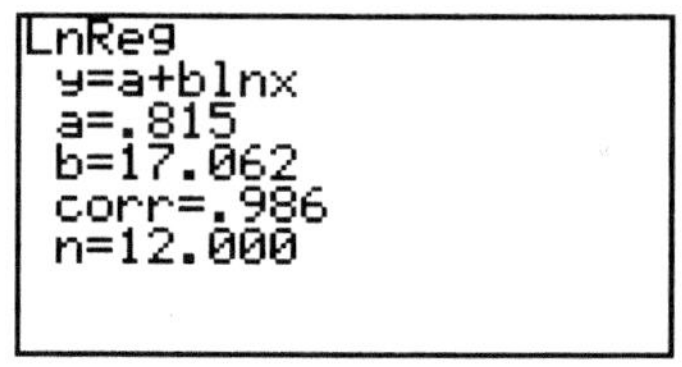

- Press [ENTER] to see the a and b of the logarithmic function. The equation is $y = 0.815 + 17.062\ln x$.
- Graph both the scatter plot and the logarithmic function on the same screen.

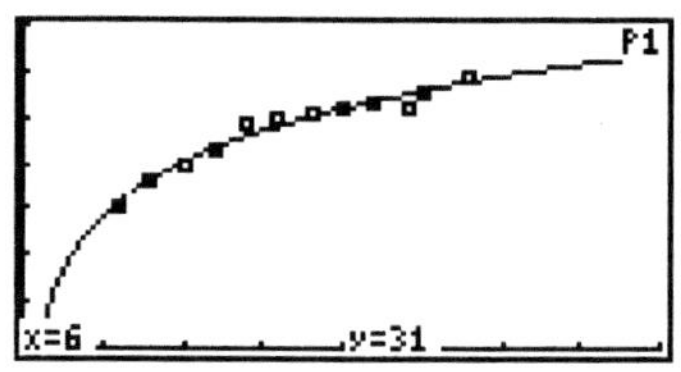

3.4 Example 5 using EXCEL

Example 5 asks you to find a logarithmic function that models the percent of moms who returned to the work force within one year after having a child (for the years from 1976 to 1998. Let x represent the number of years from 1970 and y represent percent of moms.

Create the scatter plot

- Enter labels and data for two columns in EXCEL, namely x and the percent of moms. You can use a third column for the actual year if you wish. Note that you must use x, the number of years since 1970, rather than the actual year to obtain the correct equation.
- Use the **Chart Wizard** to create the scatter plot.

Find the regression equation and put it on the scatter plot.

- Click anywhere on the scatter plot. This puts **Chart** on the menu bar.
- Go to **Chart** on the menu bar, select **Add Trendline**.
- Select Logarithmic.

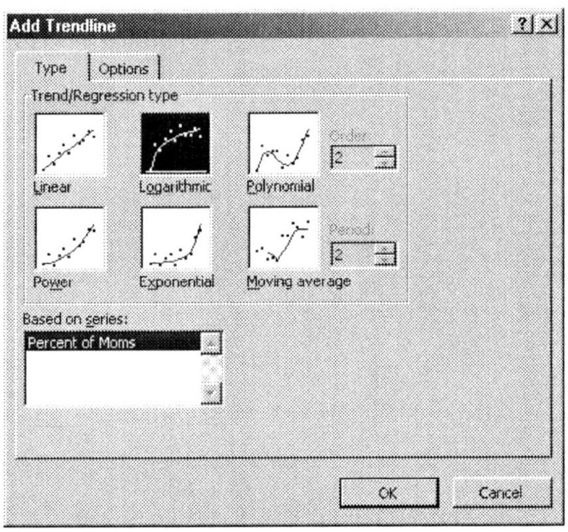

- Click on the Options tab and check Display equation on chart.
- Click **OK.**

Evaluate a Formula

- First enter a label for the first two columns, say *x* in cell B19 and *y* in cell C19.
- Enter the value of *x*, say 40 into cell B21 and the equation as $=0.815+17.062*B21$ into cell C21. Note that you can click on cell B20 to put it in the formula.
- To enter the logarithmic function you must use the function listing of EXCEL, *fn*.
- On the left of the function listing, select **Math & Trig.**
- On the right scroll down until you find **LN**. This is the logarithmic function with base *e*.
- Click **OK.** A dialog box appears. Fill it in as shown.

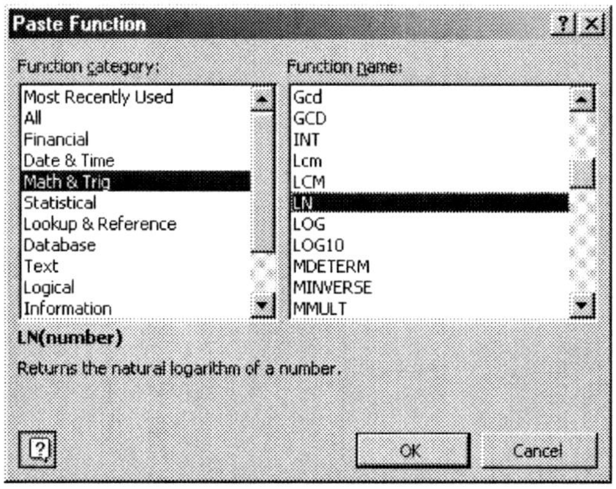

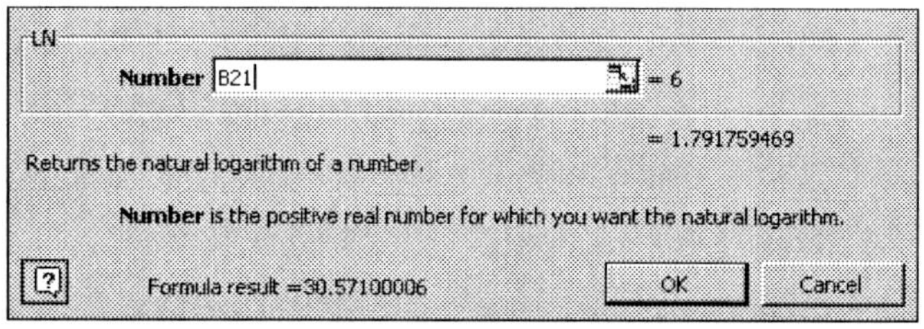

- Click **OK.**
- You will see the value 31 in cell C21 as shown on the EXCEL sheet on the next page.
- To evaluate at other values of *x,* grab the right handle of the cell with the formula, C20, and drag it down as many rows as values you wish to evaluate. Change the value of *x* in the corresponding row and press **ENTER.**

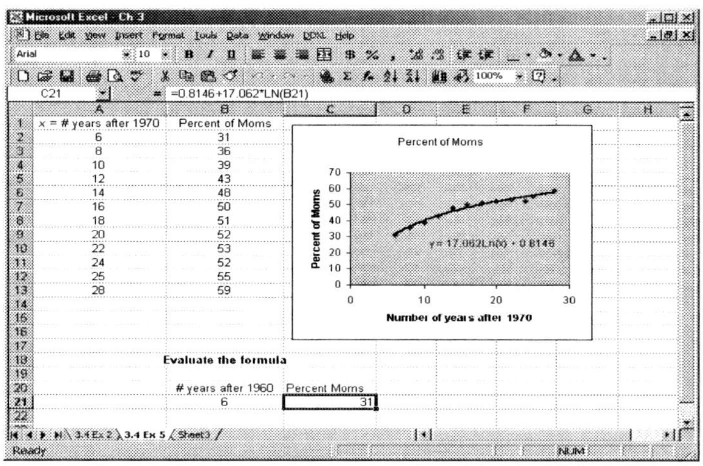

CHAPTER 4

HIGHER DEGREE POLNOMIAL AND RATIONAL FUNCTIONS

4.1 Example 1 on the TI-83, TI-83 Plus, and TI-84 PLUS

This example asks you to find the local maximum and local minimum for the cubic function $y = x^3 - 27x$.

Graph the cubic function $y = x^3 - 27x$.

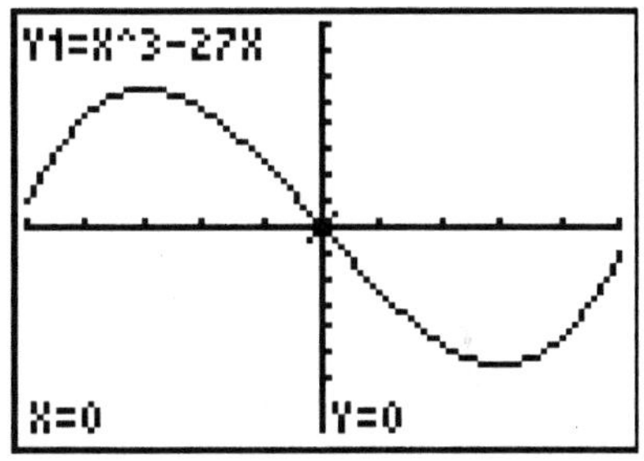

- Turn off or clear any equations or plots.
- Enter the function as Y1= $x \wedge 3 - 27x$. This is the complete graph.
- Set the window to [-5, 5, 1] by [-80, 80, 10].
- Press the [GRAPH] key and then [TRACE] to see the graph. Use the right and left arrows to move from point to point.

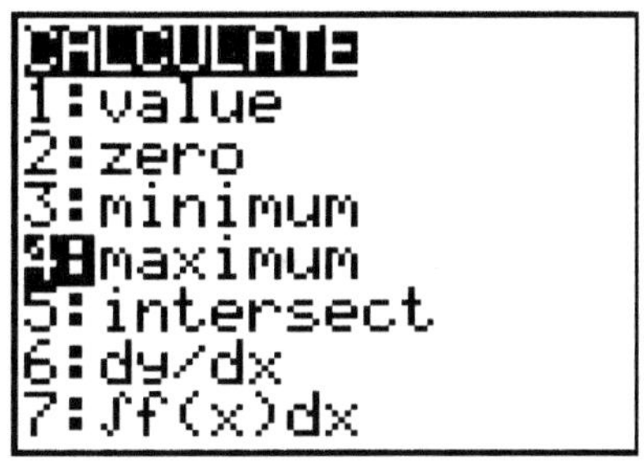

Find the local Maximum.

- To locate the local maximum on the graph, press [2nd] [TRACE] to access the CALC menu.
- Select 4: maximum.

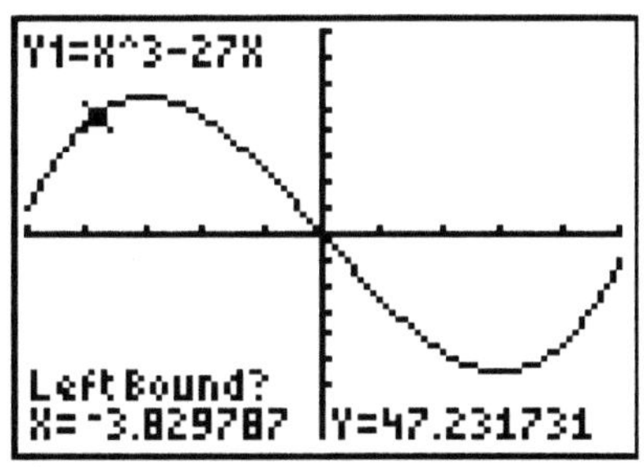

- You are back on the graph screen. The bottom of the screen is asking you to give a lower bound for the maximum you want. To do this, use the left arrow to move to a point somewhere to the left of the maximum. You could also enter a value of x here if you wish.

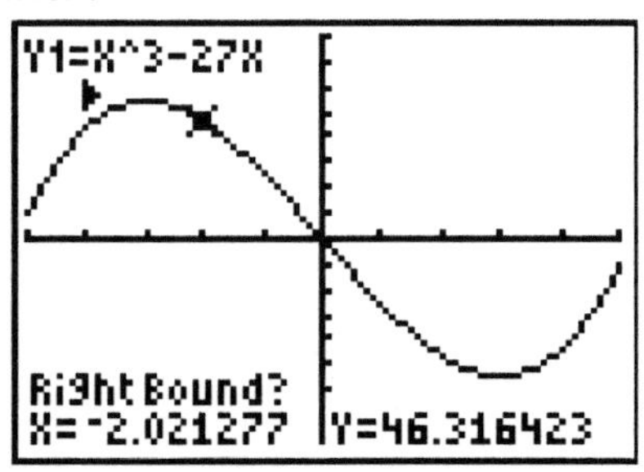

- Press [ENTER]. The bottom of the screen is now asking you to give a right bound for the maximum you want. To do this, use the right arrow to move to a point somewhere to the right of the maximum. You could also enter a value of x here if you wish.

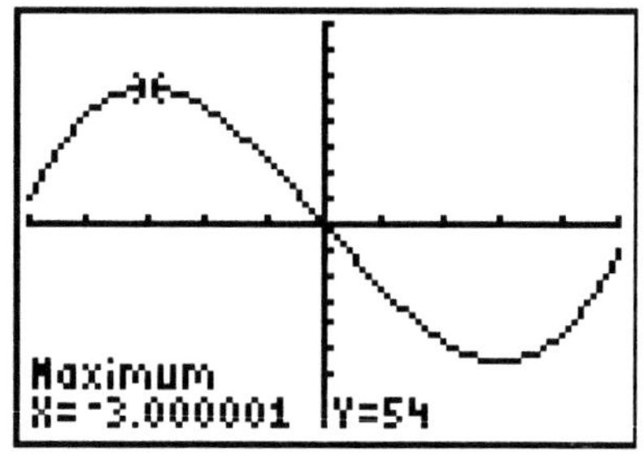

- Press [ENTER]. The bottom of the screen is now asking you for a guess. You can enter a value or just press [ENTER].
- The calculator tells you both coordinates of the maximum and moves the cursor to that point.
- Record the coordinates of the maximum point $(-3, 54)$.

Find the local Minimum.

- To locate the local minimum on the graph, press [2nd] [TRACE] to access the CALC menu.
- Select 3: minimum.

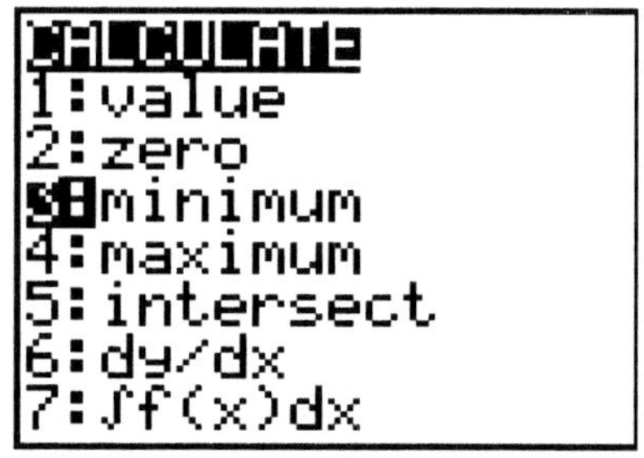

- You are back on the graph screen. The bottom of the screen is asking you to give a left bound for the minimum you want. To do this, use the right arrow to move to a point somewhere to the left of the minimum. You could also enter a value of *x* here if you wish.
- Press [ENTER]. The bottom of the screen is now asking you to give a right bound for the minimum you want. To do this, use the right arrow to move to a point somewhere to the right of the minimum. You could also enter a value of *x* here if you wish.

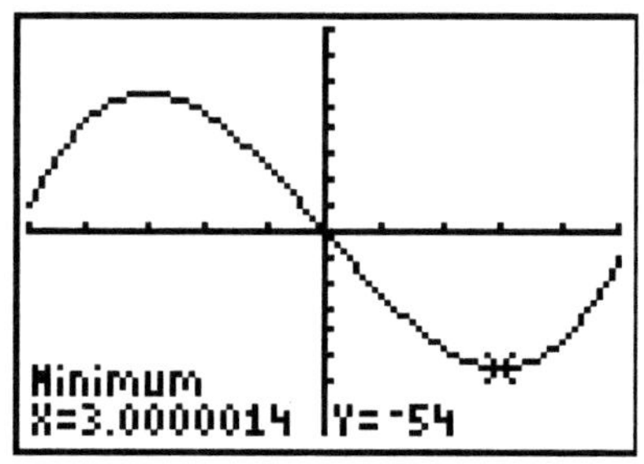

- Press [ENTER]. The bottom of the screen is now asking you for a guess. You can enter a value or just press [ENTER].
- The calculator tells you both coordinates of the minimum and moves the cursor to that point.
- Record the coordinates of the minimum point $(3, -54)$.

4.1 Example 1 on the TI-86

This example asks you to find the local maximum and local minimum for the cubic function $y = x^3 - 27x$.

Graph the cubic function $y = x^3 - 27x$.

- Turn off or clear any equations or plots.
- Enter the function as y1= $x \wedge 3 - 27x$. This is the complete graph.
- Change the window to [-5, 5, 1] by [-80, 80, 10].
- Press the [GRAPH] key and then [TRACE] to see the graph. Use the right and left arrows to move from point to point.

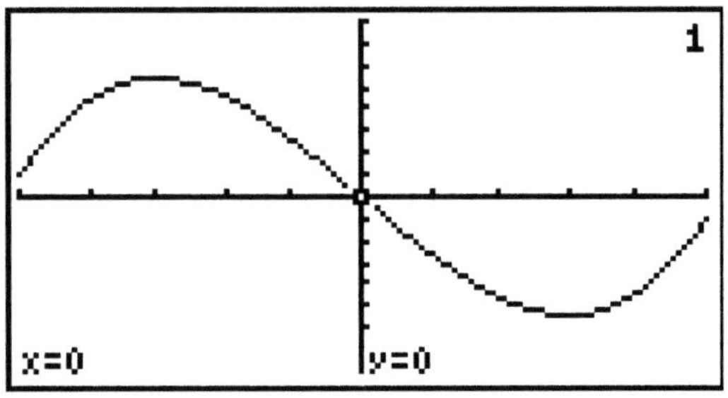

Find the local Maximum.

- To locate the local maximum on the graph, press [EXIT] [MORE] [F1] to access the MATH menu.
- Select [F5] FMAX.

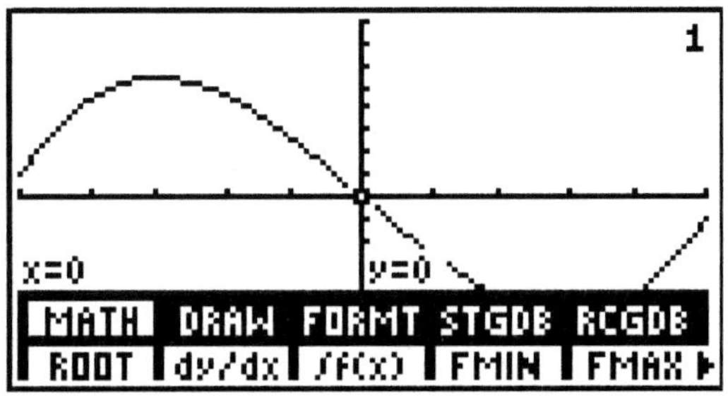

- The bottom of the screen is asking you to give a lower bound for the maximum you want. To do this, use the left arrow to move to a point somewhere to the left of the maximum. You could also enter a value of x here if you wish.

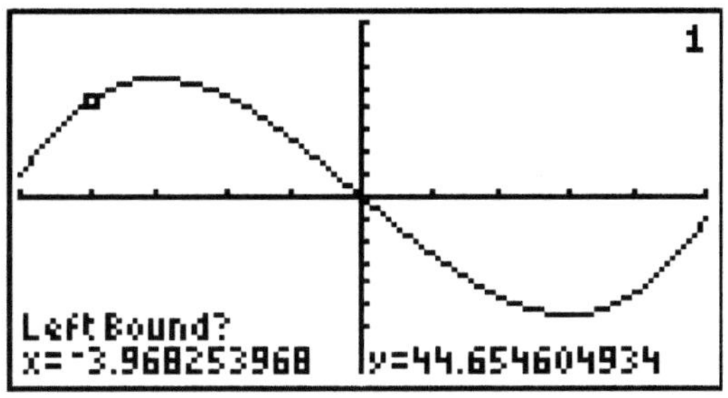

- Press [ENTER]. The bottom of the screen is now asking you to give a right bound for the maximum you want. To do this, use the right arrow to move to a point somewhere to the right of the maximum. You could also enter a value of x here if you wish.

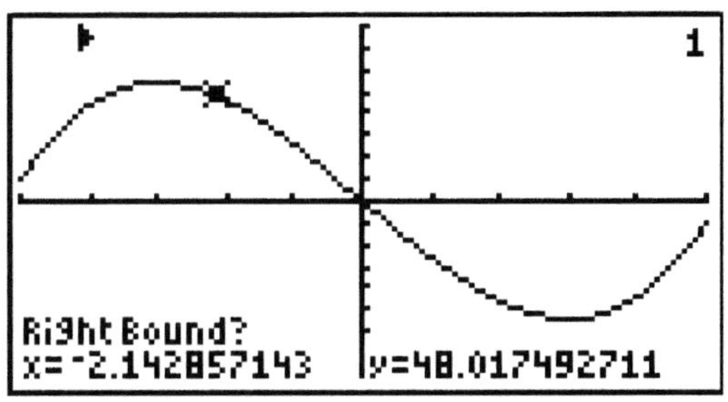

- Press [ENTER]. The bottom of the screen is now asking you for a guess. You can enter a value or just press [ENTER].
- The calculator tells you both coordinates of the maximum and moves the cursor to that point.
- Record the coordinates of the maximum point $(-3,54)$.

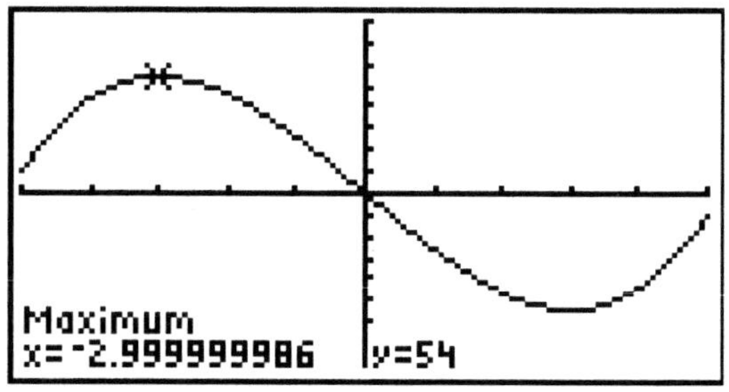

Find the local Minimum.

- To locate the local minimum on the graph, press [EXIT] [MORE] [F1] to access the MATH menu.
- Select [F4] FMIN.

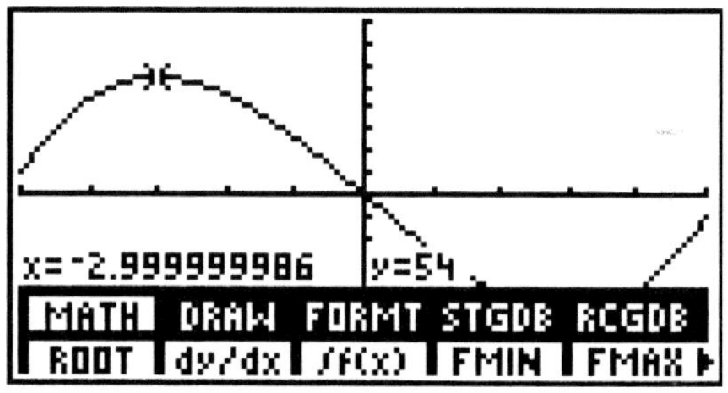

- The bottom of the screen is asking you to give a left bound for the minimum you want. To do this, use the right arrow to move to a point somewhere to the left of the minimum. You could also enter a value of x here if you wish.
- Press [ENTER]. The bottom of the screen is now asking you to give a right bound for the minimum you want. To do this, use the right arrow to move to a point somewhere to the right of the minimum. You could also enter a value of x here if you wish.
- Press [ENTER]. The bottom of the screen is now asking you for a guess. You can enter a value or just press [ENTER].
- The calculator tells you both coordinates of the minimum and moves the cursor to that point.
- Record the coordinates of the minimum point $(3,-54)$.

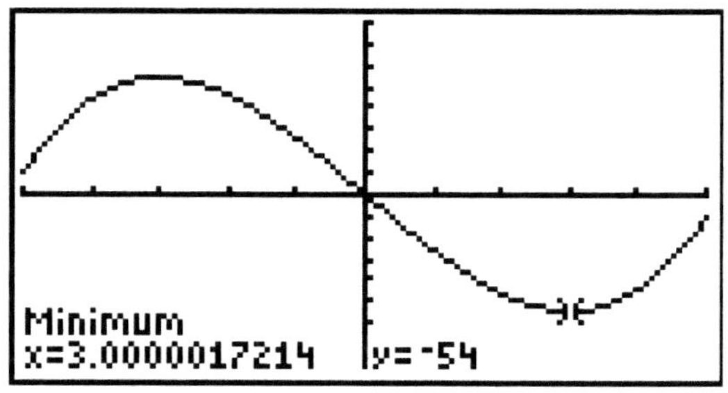

4.2 Example 2 on the TI-83, TI-83 PLUS, and TI-84 PLUS

This example asks you to find a linear function and a cubic function that models the money spent by domestic travelers in the U.S. for each of the years 1987-96. Let x represent the number of years from 1980 and y represent dollars spent in billions.

Create the scatter plot

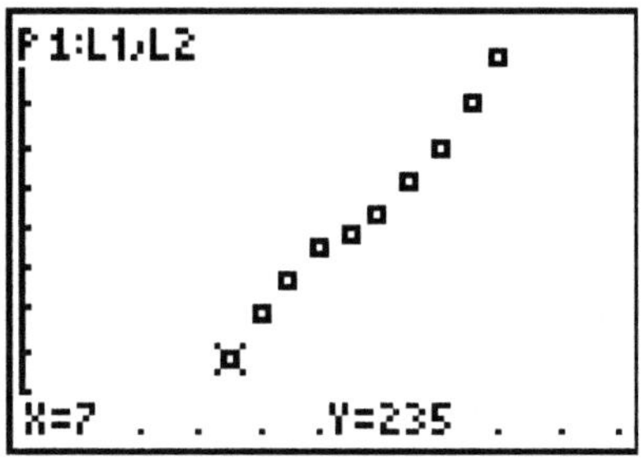

- Enter the data into lists L1 and L2. Note that $x = 0$ is the year 1980, so the first data point has $x = 7$ for the year 1987.
- Examine the data to set the Window at [0, 20, 2] by [200,400, 20].
- Create the scatter plot of the data.

Find and graph the linear function model for this data

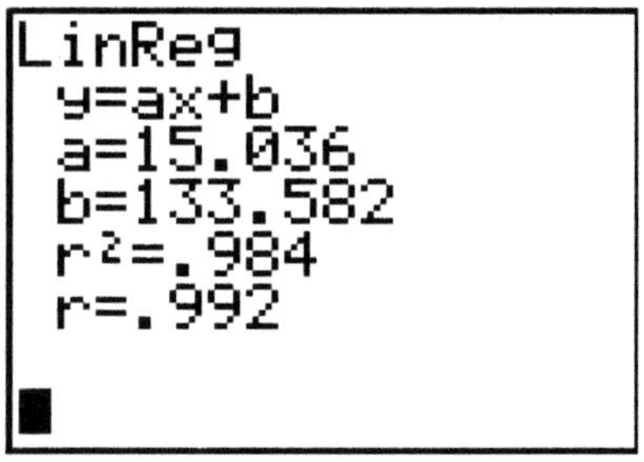

- Go to the STAT CALC menu.
- Select 4:LinReg(ax+b).
- Enter the names of the lists L1 and L2 that contain the data.
- Press ENTER to see the a and b of the linear regression model. The equation is $y = 15.036x + 133.582$.

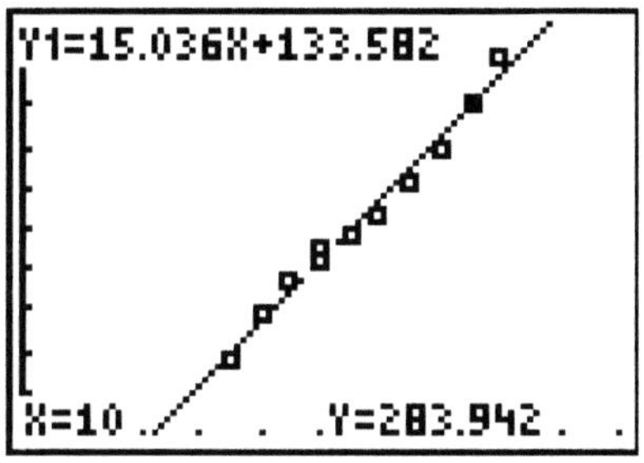

- Put the regression equation into the Y= menu.
- Place the cursor on Y1.
- Press VARS, select 5:Statistics EQ 1:RegEQ.
- Press ENTER to paste the equation as Y1.
- Graph both the scatter plot and the linear function on the same screen.

Find and graph the cubic function model for this data

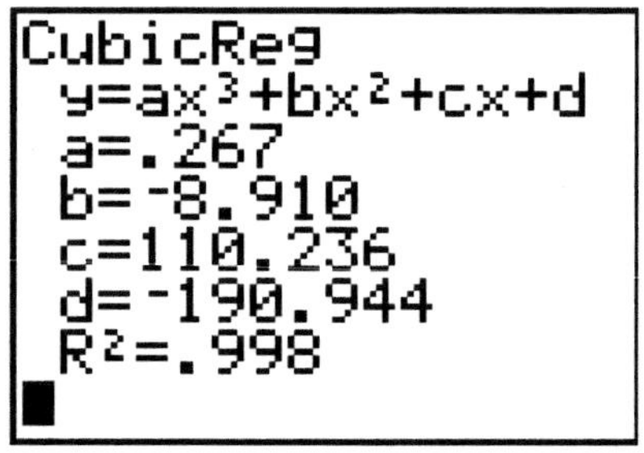

- Go to the STAT CALC menu.
- Select 6:CubicReg.
- Enter the names of the lists that contain the data.
- Press [ENTER] to see the a, b, and c of the cubic model. The equation is $y = 0.267x^3 - 8.910x^2 + 110.236x - 190.944$.

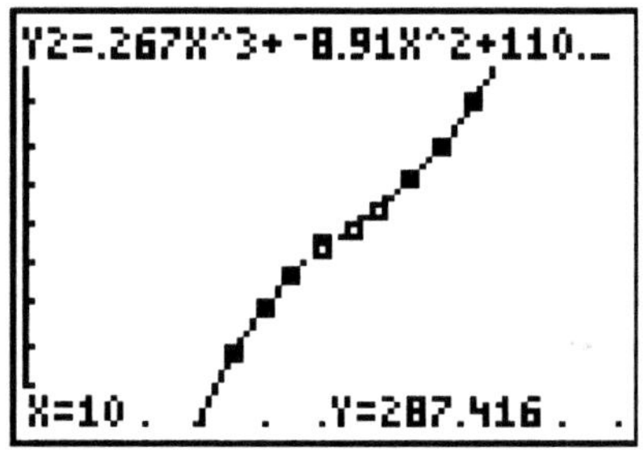

- Turn off Y1.
- Put the regression equation into the Y= menu. Place the cursor on Y2. Press [VARS], select 5:Statistics EQ 1:RegEQ.
- Press [ENTER] to paste the equation as Y2.
- Graph both the scatter plot and the cubic function on the same screen.
- The graphs indicate that the cubic function is a better model for this data.

4.2 Example 2 on the TI-86

Example 2 asks you to find a linear function and a cubic function that models the money spent by domestic travelers in the U.S. for each of the years 1987-96. Let x represent the number of years from 1980 and y represent dollars spent in billions.

Create the scatter plot

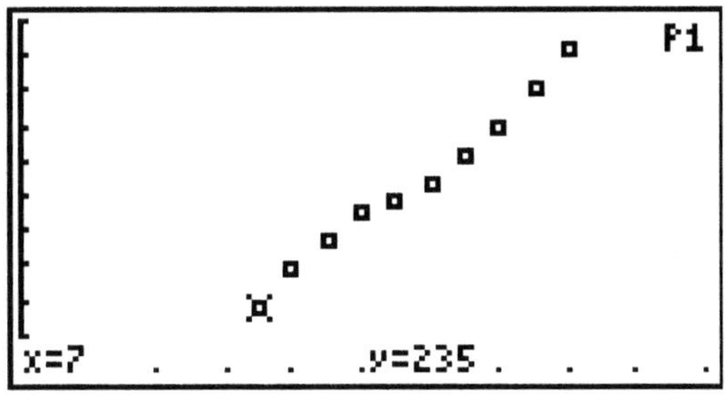

- Enter the data into lists L1 and L2. Note that $x = 0$ is the year 1980, so the first data point has $x = 7$ for the year 1987.
- Examine the data to set the Window at [0, 20, 2] by [200, 400, 20].
- Create the scatter plot of the data.

Find and graph the linear function model for this data

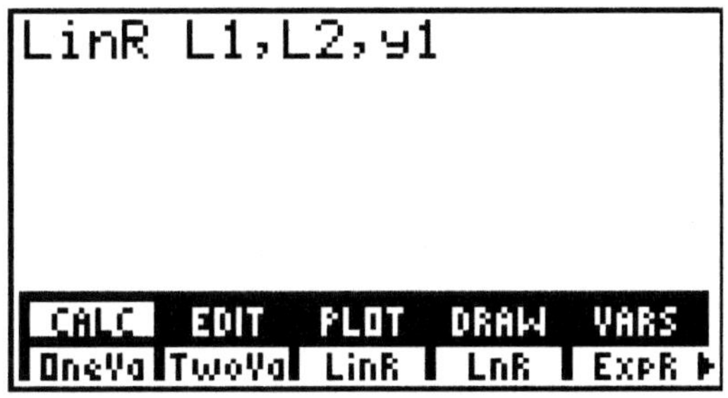

- Go to the STAT CALC menu.
- Press [F3] to select LinR, the linear function.
- Enter the names of the lists L1 and L2 that contain the data, and the location of a y = function. This will allow us to graph both the linear function model and the scatter plot on the same screen.
- Press [ENTER] to see the a and b of the linear function. The equation is $y = 15.036x + 133.582$.
- Graph both the scatter plot and the linear function on the same screen.

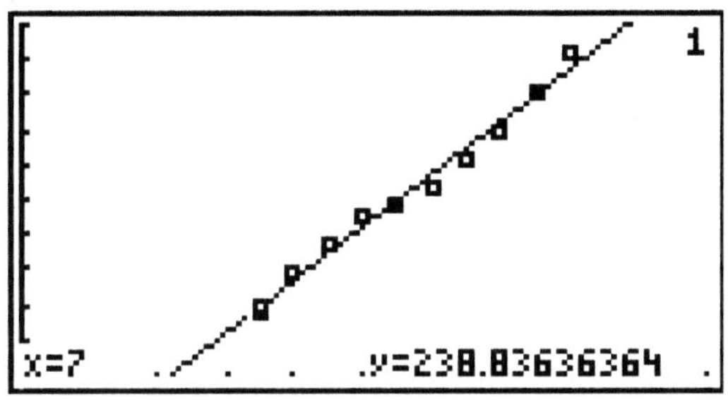

Find and graph the cubic function model for this data

- Go to the STAT CALC menu.
- Press [MORE] once then [F5] to select P3Reg, the cubic function.
- Enter the names of the lists that contain the data, and the location y2 to paste the function into y2.

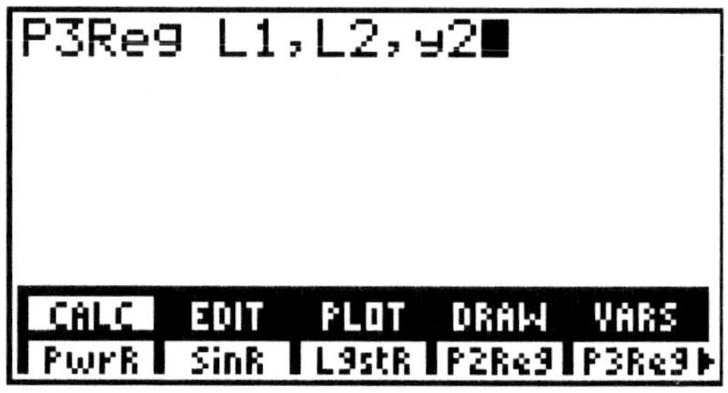

- Press [ENTER] to see the a, b, c, and d of the cubic function. The equation is $y = 0.267x^3 - 8.910x^2 + 110.236x - 190.944$.
- Use the right arrow to scroll through all of the values.
- Turn off y1. The cubic regression equation is already pasted into y2.
- Graph both the scatter plot and the cubic function on the same screen.
- The graphs indicate that the cubic function is a better model for this data.

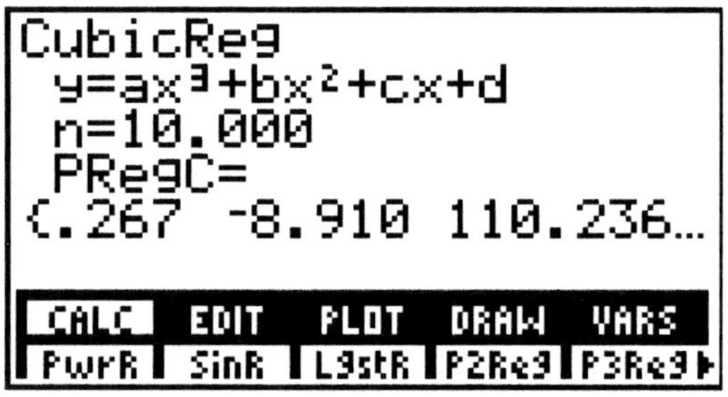

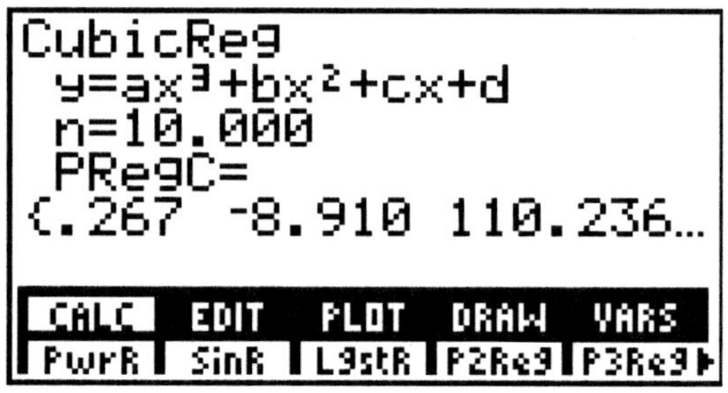

4.2 Example 2 using EXCEL

This example asks you to find a linear function and a cubic function that models the money spent by domestic travelers in the U.S. for each of the years 1987-96. Let x represent the number of years from 1980 and y represent dollars spent in billions.

Create the scatter plot

- Enter labels and data for two columns in EXCEL, namely x and the dollars spent in billions. You can use a third column for the actual year if you wish. Note that you must use x, the number of years since 1980, rather than the actual year to obtain the correct equation.
- Use the **Chart Wizard** to create the scatter plot.

Find the regression equation and put it on the scatter plot.

- Click anywhere on the scatter plot. This puts **Chart** on the menu bar.
- Go to **Chart** on the menu bar, select **Add Trendline**.
- Select **Polynomial**. Select **Order 3**.
- Click on the **Options** tab and check **Display equation on chart**.
- Click **OK.**

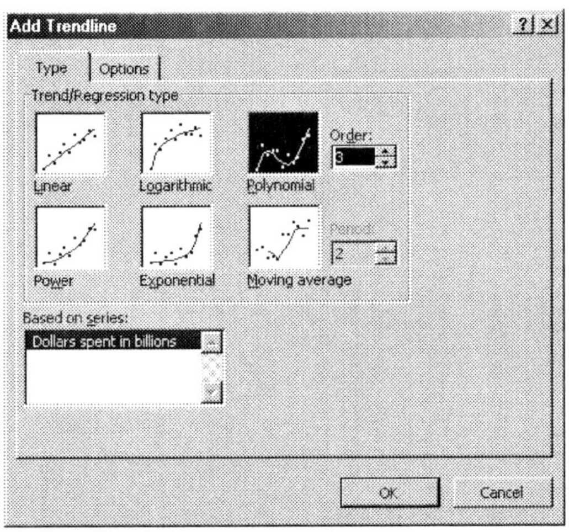

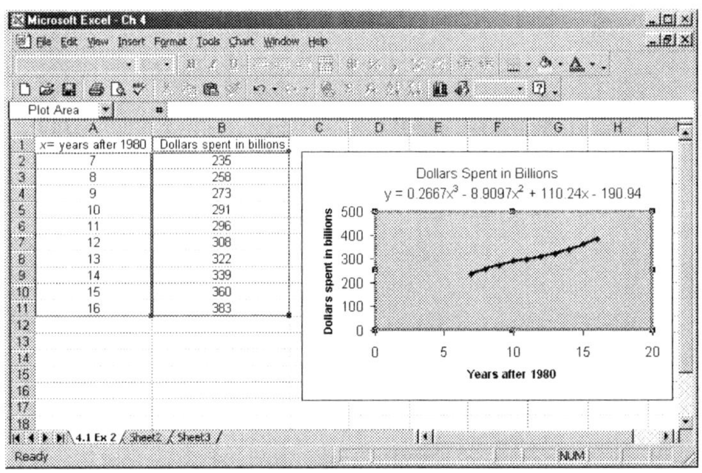

- Repeat these steps for the linear case, select **Add Trendline linear**.

CHAPTER 5

SYSTEMS OF EQUATIONS AND MATRICES

5.2 Example 2 on the TI-83, TI-83 Plus, and TI-84 PLUS

This example asks you to solve the following system of linear equations with matrices.

$$x + y + z = 150{,}000$$
$$0.10x + 0.08y + 0.07z = 12{,}900$$
$$x - y - z = -20{,}000$$

First represent the system as the augmented matrix $\begin{bmatrix} 1 & 1 & 1 & 150000 \\ 0.10 & 0.08 & 0.07 & 12900 \\ 1 & -1 & -1 & -20000 \end{bmatrix}$.

Solve the System of Linear Equations with Matrices

Enter the Matrix

- Set the MODE to the word FLOAT.
- Press [2nd] [x^{-1}] to access the MATRIX menu.
- Use the right arrow to go to EDIT.

Move the cursor to 1: [A] and press [ENTER]. Note that if you used this matrix name before, it will have a dimension next to it.

- Enter the dimension of matrix A as 3 x 4.
- Enter the values into the matrix as shown. Press [ENTER] after each entry.
- Note that the position is given at the bottom of the screen as 1,1=2 etc.
- This matrix will need two screens. Use the right arrow to see last column.
- Press [2nd] [MODE] to QUIT and return to the Home screen.

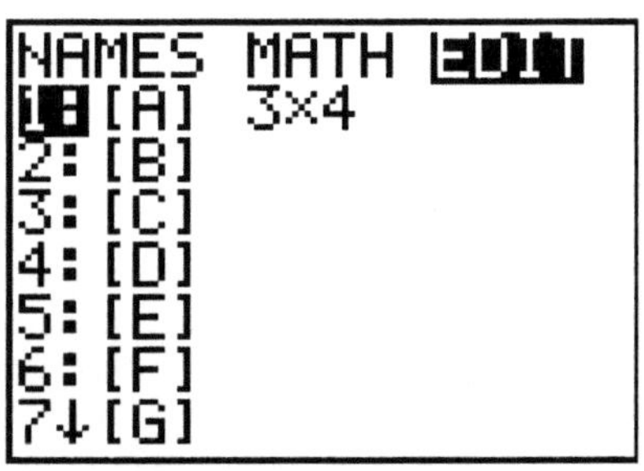

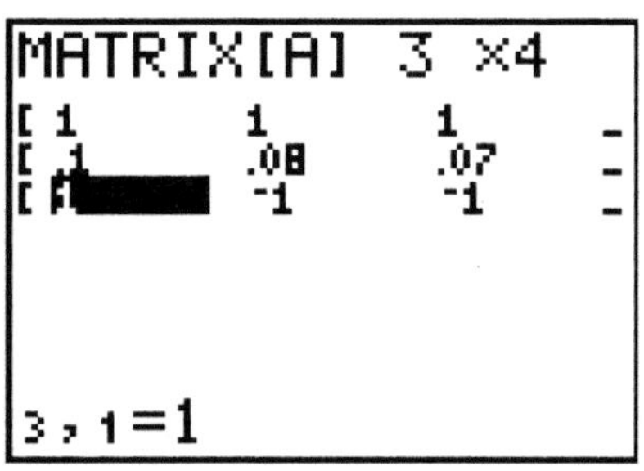

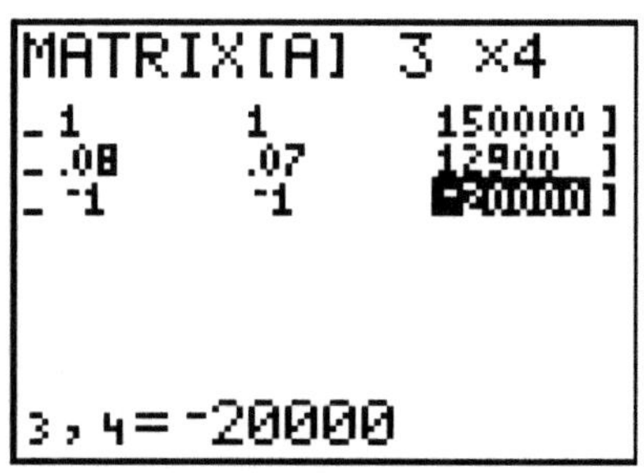

View the Matrix on the Home Screen

- Press [2nd] [x^{-1}] to access the MATRIX menu. You are in the NAMES menu.
- Move the cursor to 1: [A] and press [ENTER]. This will put [A] on the Home screen.
- Press [ENTER] to view the matrix on the Home screen.
- Use the right arrow to scroll through the entire matrix.

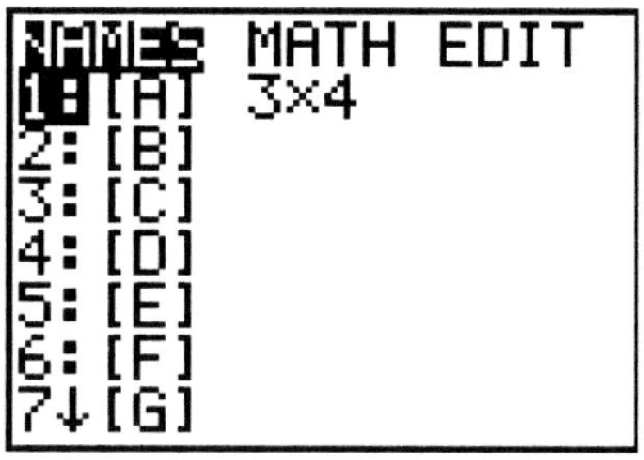

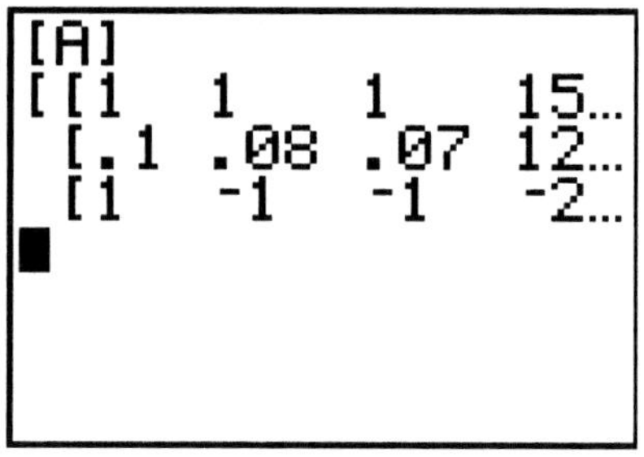

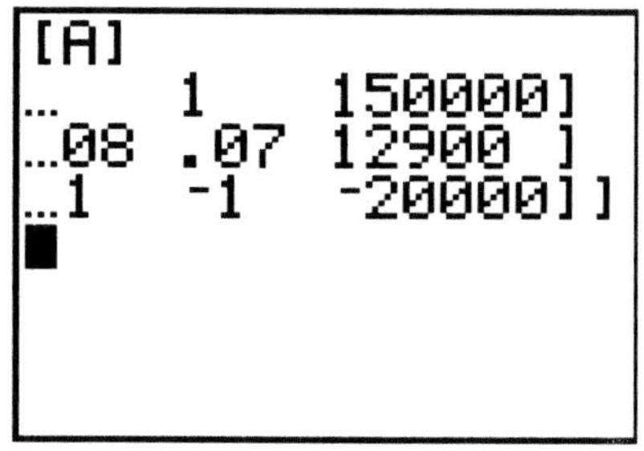

Find the reduced row-echelon form of the matrix, rref ([A])

- Press [2nd] [x^{-1}] to access the MATRIX menu.
- Use the right arrow to go to MATH.
- Use the down arrow to select B: rref(.
- Press [ENTER]. This puts rref(on the Home screen.

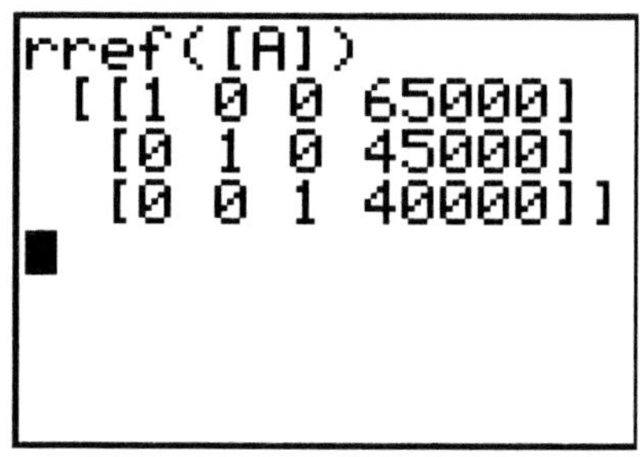

- Enter the name of the matrix you wish to use in the parentheses.
- You must use MATRIX NAMES to enter the name of the matrix. Otherwise you get an ERR: DATA TYPE message.
- To enter the matrix name, press [2nd] [x^{-1}] to access the MATRIX menu. You are in MATRIX NAMES.
- Select 1:[A]
- Press [ENTER]. Use the right arrow to see the entire matrix.

5.2 Example 2 on the TI-86

This example asks you to solve the following system of linear equations with matrices.

$$x+y+z=150{,}000$$
$$0.10x+0.08y+0.07z=12{,}900$$
$$x-y-z=-20{,}000$$

First represent the system as the augmented matrix $\begin{bmatrix} 1 & 1 & 1 & 150000 \\ 0.10 & 0.08 & 0.07 & 12900 \\ 1 & -1 & -1 & -20000 \end{bmatrix}$.

Solve the System of Linear Equations with Matrices

Enter the Matrix

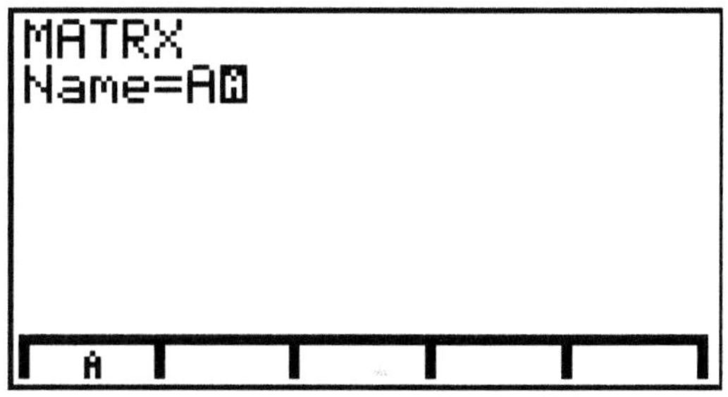

- Press [2nd] [7] to access the MATRIX menu.
- Press [F2] to go to EDIT.
- You are asked to give the matrix a name. Notice that the cursor is a flashing A , which means that you are already in ALPHA mode. Use [LOG] to name it as matrix A.
- Once you have named a matrix, its name appears in one of the boxes at the bottom of the screen.
- Press [ENTER]. Enter the dimension of matrix A as 3 x 4. Enter the values into the matrix as shown. Note that the position is given at the bottom of the screen as 1,1=2 etc. This matrix will need two screens. Use the right arrow to see last column. When you enter the fraction in fraction form, the calculator shows the decimal equivalent.
- Press [EXIT] to return to the Home screen.

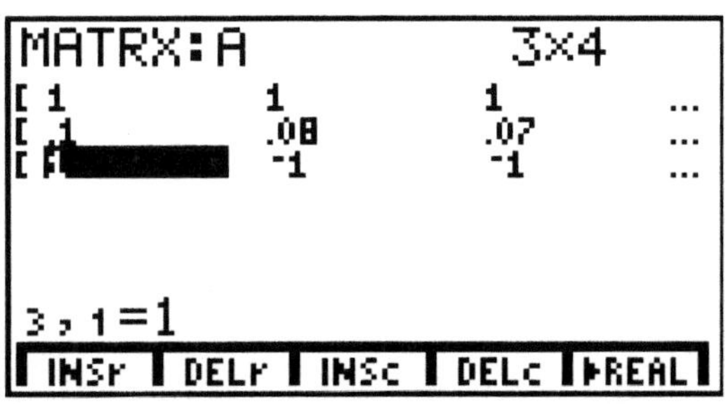

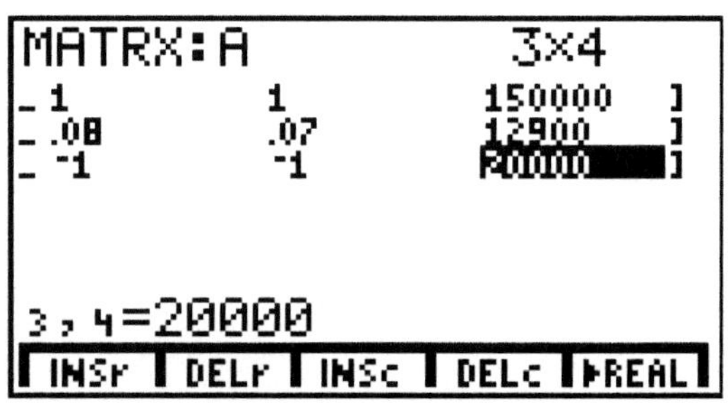

View the Matrix on the Home Screen

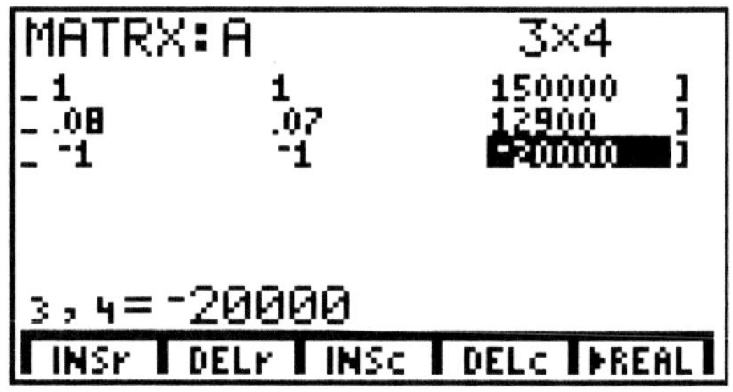

- Press [2nd] [7] to access the MATRIX menu.
- Press [F1] to go to NAMES.
- Press [F1] to select matrix A. This matrix may be stored in a different box.
- Press [ENTER] to view the matrix on the Home screen.
- Press [EXIT] to return to the Matrix menu.

Find the reduced row-echelon form of the matrix, rref ([A])

- If you are on the Home screen, press [2nd] [7] to access the MATRIX menu.
- Press [F4] OPS.
- Press [F5] to select rref.
- Press [ENTER]. This puts rref on the Home screen.
- Enter the name of the matrix you wish to use in the parentheses. You may use MATRIX NAMES to enter the name of the matrix. You may also use [ALPHA] [LOG] to enter the name of the matrix as A.
- Press [ENTER] to see the row-echelon form of matrix A.

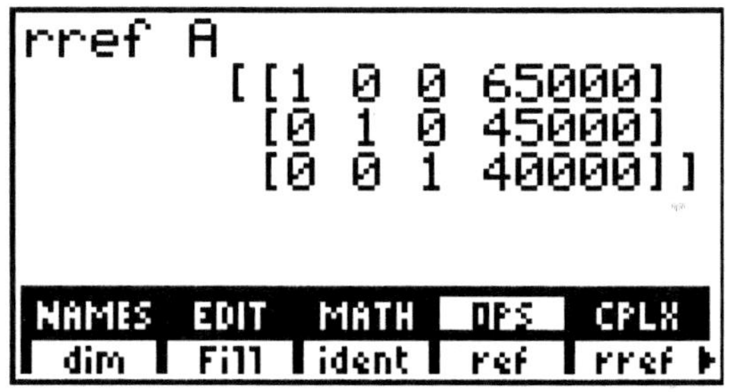

5.2 Example 2 using EXCEL

Solve the System of Linear Equations with Matrices

You must use row operations of the text to solve this system in EXCEL. This is a very tedious process. Let's wait and solve this using the matrix inverse of A that is studied in section 5.5.

5.3 Example 2b on the TI-83, TI-83 PLUS, and TI-84 PLUS

Example 2b asks you to perform some matrix arithmetic using addition and subtraction.

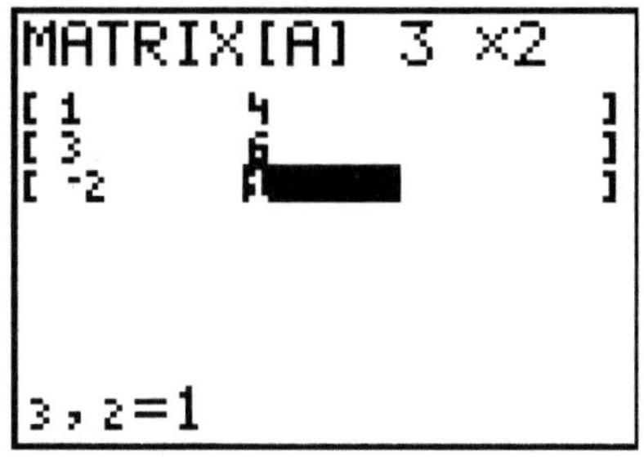

Find $\begin{bmatrix} 1 & 4 \\ 3 & 6 \\ -2 & 1 \end{bmatrix} - \begin{bmatrix} -5 & 2 \\ 3 & 5 \\ 12 & 3 \end{bmatrix} + \begin{bmatrix} 2 & 3 \\ -4 & -8 \\ 2 & 0 \end{bmatrix}$.

Perform Matrix Arithmetic

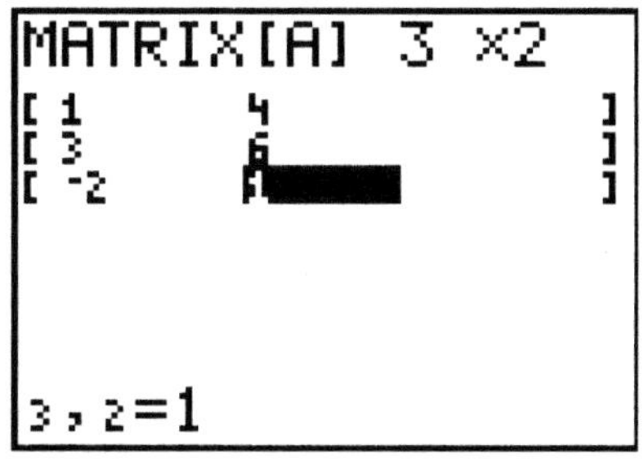

Enter the Matrices as A, B, and C

- Press [2nd] [x^{-1}] to access the MATRIX menu.
- Use the right arrow to go to EDIT.
- Move the cursor to 1: [A] and press [ENTER]. Note that if you used this matrix name before, it will have a dimension next to it.
- Enter the dimension of matrix A as 3 x 2. Enter the values into the matrix as shown. Note that the position is given at the bottom of the screen as 1,1=1 etc.
- Press [2nd] [MODE] to QUIT and return to the Home screen.
- Repeat the process to enter matrices B and C.
- Press [2nd] [MODE] to QUIT and return to the Home screen.

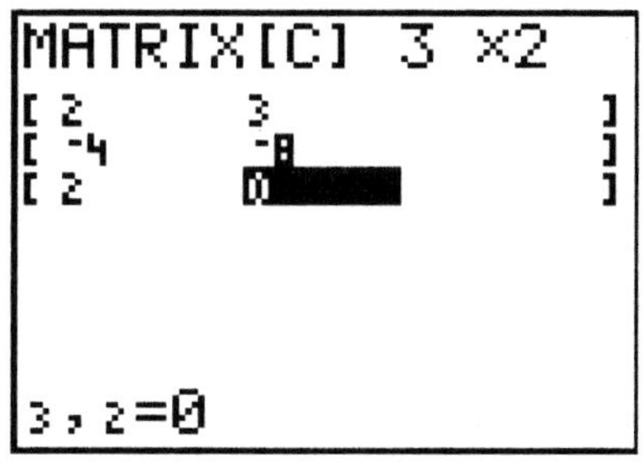

Perform the Matrix Arithmetic on the Home Screen

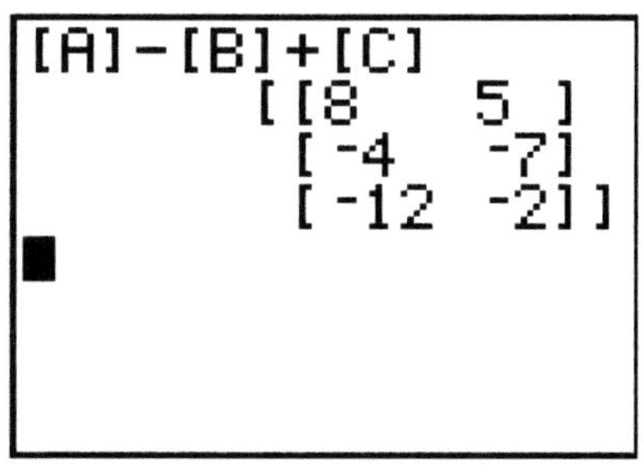

- Press [2nd] [MODE] to QUIT and return to the Home screen.
- You must use MATRIX NAMES to enter the names of the matrices.
- Press [2nd] [x^{-1}] to access MATRIX NAMES and select 1: [A].
- Use the operation keys to add and subtract as shown.
- Repeat the process to select matrices 2:[B] and 3:[C].

Multiply a matrix by a number

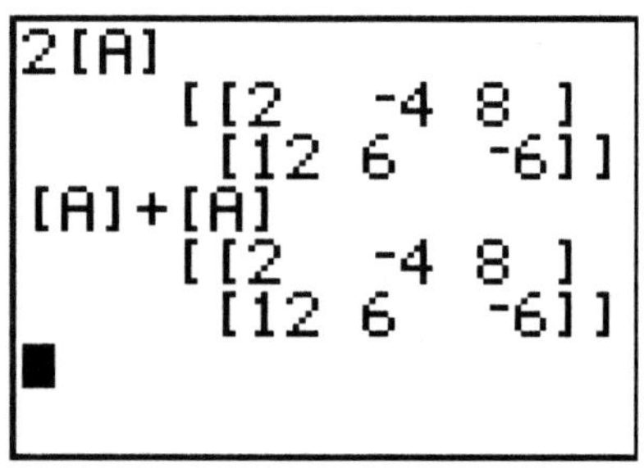

Find $2A = 2\begin{bmatrix} 1 & -2 & 4 \\ 6 & 3 & -3 \end{bmatrix}$ on page 479.

- Enter the new matrix [A]. On the Home screen enter [2] [A] or [A] + [A] as shown. Press [ENTER] to see the result.

5.3 Example 2b on the TI-86

Example 2b asks you to perform some matrix arithmetic using addition and subtraction.

Find $\begin{bmatrix} 1 & 4 \\ 3 & 6 \\ -2 & 1 \end{bmatrix} - \begin{bmatrix} -5 & 2 \\ 3 & 5 \\ 12 & 3 \end{bmatrix} + \begin{bmatrix} 2 & 3 \\ -4 & -8 \\ 2 & 0 \end{bmatrix}$.

Perform Matrix Arithmetic

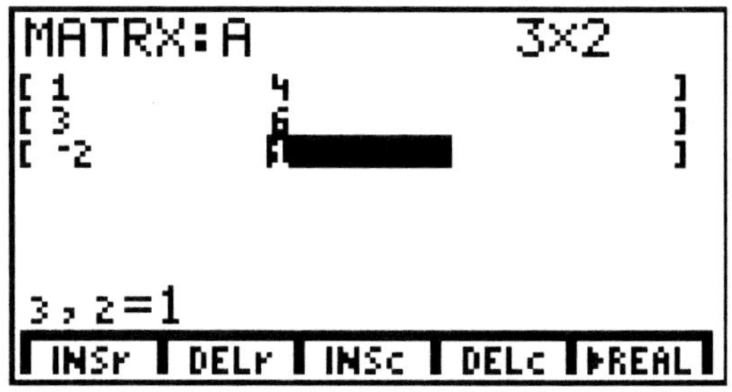

Enter the Matrices as A, B, and C

- Press [2nd] [7] to access the MATRIX menu.
- Press [F2] to go to EDIT.
- You are asked to give the matrix a name. Notice that the cursor is a flashing A, which means that you are already in ALPHA mode. Use the [LOG] key to name it as matrix A.

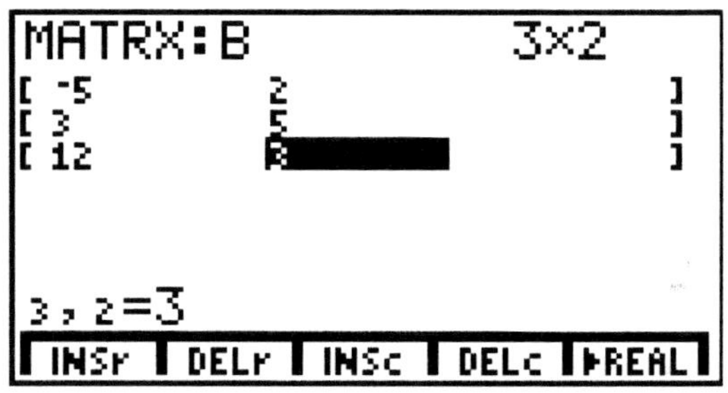

- Press [ENTER].
- Enter the dimension of matrix A as 3 x 2. Enter the values into the matrix as shown. Note that the position is given at the bottom of the screen as 1,1=1 etc.

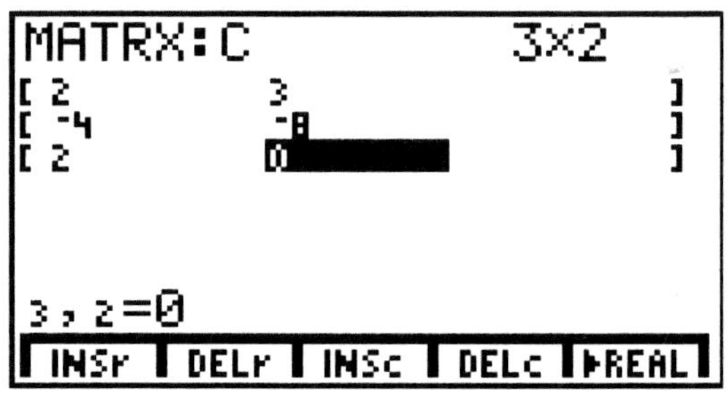

- Press [EXIT] to return to the Home screen.
- Repeat the process to enter matrices B and C, i.e. Press [2nd] [7] to return to the MATRIX menu etc.
- Press [EXIT] to return to the Home screen.

Perform the Matrix Arithmetic on the Home Screen

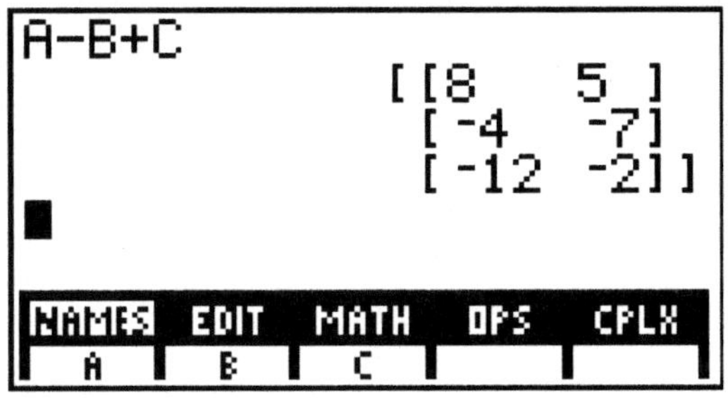

- Press [EXIT] to return to the Home screen.
- Use MATRIX NAMES to enter the names of the matrices. You may also use [ALPHA] [LOG] to enter the name of the matrix as A, etc.

Multiply a matrix by a number

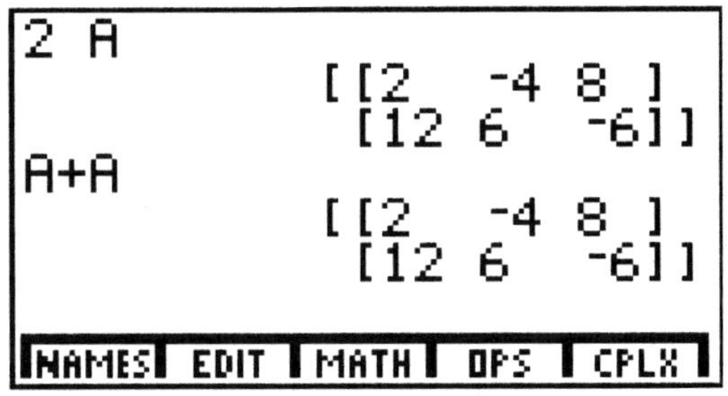

Find $2A = 2\begin{bmatrix} 1 & -2 & 4 \\ 6 & 3 & -3 \end{bmatrix}$ on page 848.

- Use MATRIX EDIT to enter the new matrix as A.
- On the Home screen enter as 2 [A] or [A]+ [A] as shown.
- Press [ENTER] to see the result.

5.3 Example 2b using EXCEL

Example 2b asks you to perform some matrix arithmetic using addition and subtraction.

Find $\begin{bmatrix} 1 & 4 \\ 3 & 6 \\ -2 & 1 \end{bmatrix} - \begin{bmatrix} -5 & 2 \\ 3 & 5 \\ 12 & 3 \end{bmatrix} + \begin{bmatrix} 2 & 3 \\ -4 & -8 \\ 2 & 0 \end{bmatrix}$.

Perform Matrix Arithmetic

Enter the Matrices as A, B, and C

- Enter each matrix, leaving at least one column between them.
- Click in the first cell where you wish to see the resulting matrix.
- Enter the expression as a function.
- In cell A6 put =A1:B3 – E1:F3 + I1:J3. To do this, highlight the range of cells that make up the first matrix, enter the operation sign + or - , then highlight the range of cells of the second matrix, etc.
- Press **CNTRL SHIFT ENTER** on the keyboard to see the resulting matrix.
- Grab the right handle and drag through a 3x2 matrix for the result.

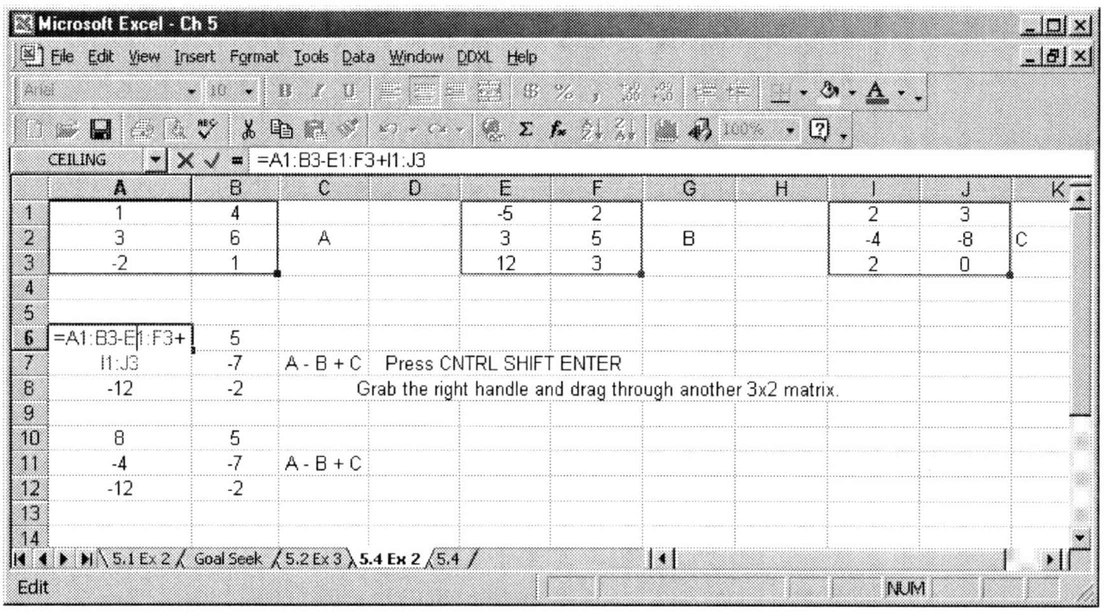

Multiply a matrix by a number

Find $2A = 2\begin{bmatrix} 1 & -2 & 4 \\ 6 & 3 & -3 \end{bmatrix}$ on page 479.

- Enter the matrix.
- Highlight the range of cells for the result.
- In the first cell of the result enter the equation = 2*(range of cells for A)
- Enter the expression as a function.
- In cell A4 put =2*A1:C2. To do this, highlight the range of cells that make up the matrix.
- Press **CNTRL SHIFT ENTER** on the keyboard to see the resulting matrix.
- Grab the right handle and drag through a 3x2 matrix for the result.

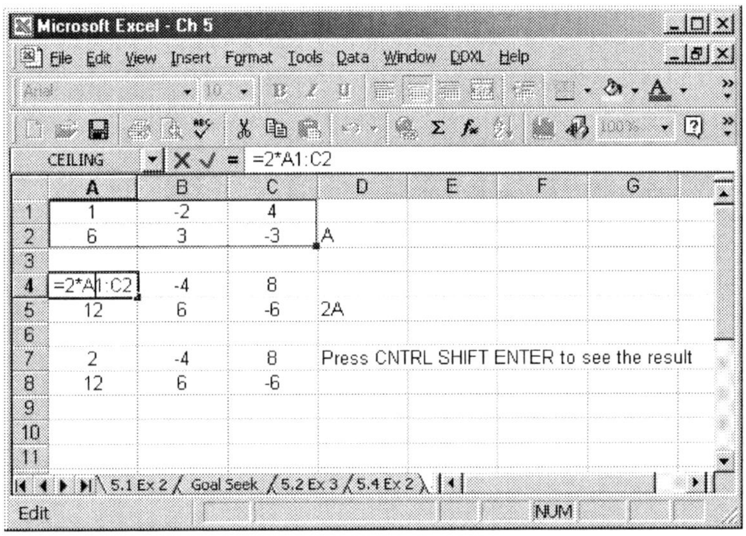

5.3 Example 6 on the TI-83, TI-83 PLUS, and TI-84 PLUS

This example asks you to multiply two matrices. Note that the number columns of the first matrix must be the same as the number of rows of the second. So we multiply BA .

$$B=\begin{bmatrix}3 & 12 & 15\\ 5 & 16 & 10\\ 10 & 11 & 6\end{bmatrix},\ A=\begin{bmatrix}12 & 10\\ 10 & 8\\ 5 & 9\end{bmatrix}$$

Perform Matrix Multiplication

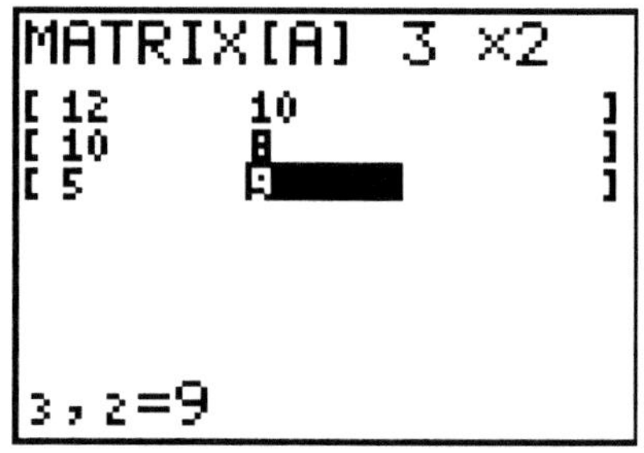

Enter the Matrices

- Press [2nd] [x^{-1}] to access the MATRIX menu.
- Use the right arrow to go to EDIT.
- Move the cursor to 1: [A] and press [ENTER]. Note that if you used this matrix name before, it will have a dimension next to it.
- Enter the dimension of matrix A as 3 x 2. Enter the values into the matrix as shown. Note that the position is given at the bottom of the screen as 1,1=12 etc.

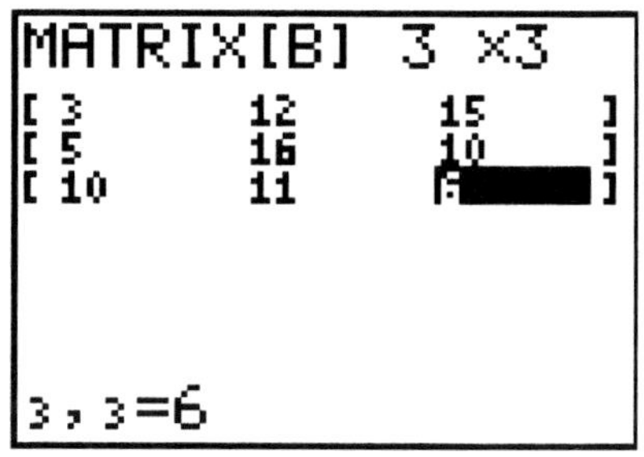

- Repeat the process to enter matrix B, i.e. Press [2nd] [x^{-1}] to return to the MATRIX menu etc.
- Press [2nd] [MODE] to QUIT and return to the Home screen.

Display a matrix on the Home Screen

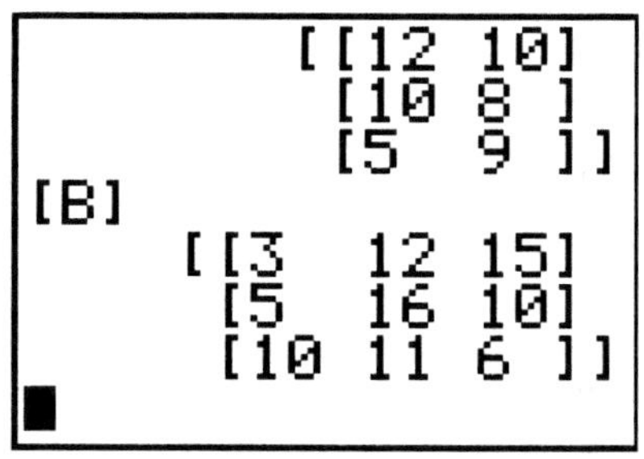

- Press [2nd] [MODE] to QUIT and return to the Home screen.
- Press [2nd] [x^{-1}] to access the MATRIX menu.
- In MATRIX NAMES select 1:[A]. Press [ENTER].
- Repeat and select 2:[B].

Perform the Matrix Multiplication on the Home Screen

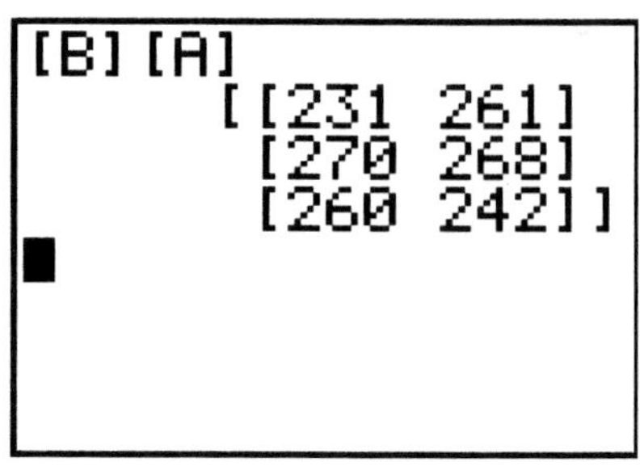

- Use MATRIX NAMES to enter the names of the matrices. You may use the multiplication key to multiply the matrices as shown, but it not necessary.
- Press [ENTER] to see the product matrix.
- Note that you get a dimension error when you try to do [A] [B].

5.3 Example 6 on the TI-86

This example asks you to multiply two matrices. Note that the number columns of the first matrix must be the same as the number of rows of the second. So we multiply BA.

$$B=\begin{bmatrix}3 & 12 & 15\\ 5 & 16 & 10\\ 10 & 11 & 6\end{bmatrix},\ A=\begin{bmatrix}12 & 10\\ 10 & 8\\ 5 & 9\end{bmatrix}$$

Perform Matrix Multiplication

Enter the Matrices

- Press 2nd 7 to access the MATRIX menu.
- Press F2 to go to EDIT. You are asked to give the matrix a name. If you already used A for the name of a matrix, press the f key below it. Otherwise, use the LOG key to name it as matrix A.
- Press ENTER. Enter the dimension of matrix A as 3 x 2. Enter the values into the matrix as shown. Note that the position is given at the bottom of the screen as 1,1=12 etc.

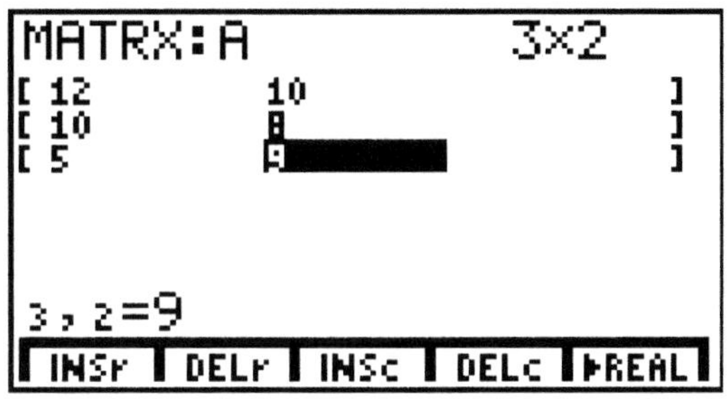

- Repeat the process to enter matrix B, i.e. Press 2nd 7 to return to the MATRIX menu etc.
- Press EXIT to return to the Home screen.

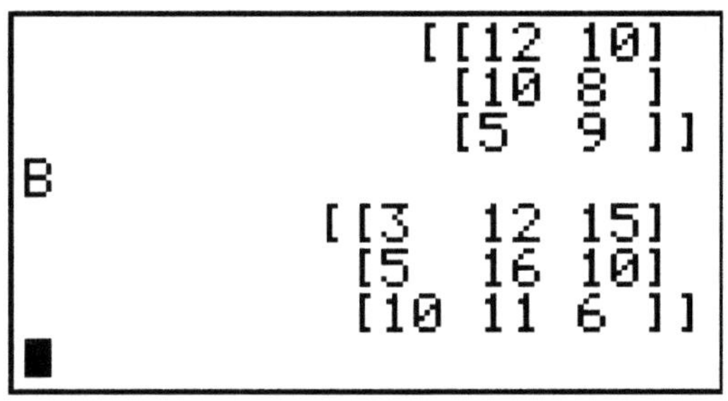

Display a matrix on the Home Screen

- Press EXIT to return to the Home screen.
- Press 2nd 7 to access the MATRIX menu.
- Press the F key that contains matrix [A]. Press ENTER.
- Repeat and select 2:[B].

Perform the Matrix Multiplication

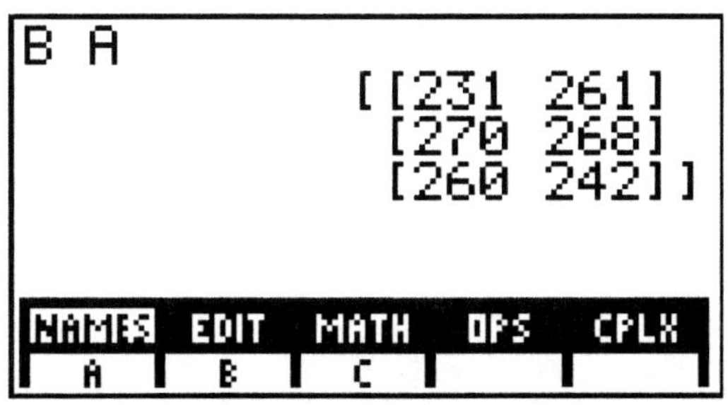

- Press EXIT to return to the Home screen.
- Use MATRIX NAMES to enter the names of the matrices. You may also use ALPHA SIN to enter the name of the matrix as B, and use MATRIX NAMES or ALPHA LOG to enter the name of matrix A.
- Press ENTER to see the product matrix.

5.3 Example 6 Using EXCEL

Example 6 – Advertising – asks you to multiply two matrices. Note that the number columns of the first matrix must be the same as the number of rows of the second. So we multiply BA.

$$B=\begin{bmatrix}3 & 12 & 15\\ 5 & 16 & 10\\ 10 & 11 & 6\end{bmatrix},\ A=\begin{bmatrix}12 & 10\\ 10 & 8\\ 5 & 9\end{bmatrix}$$

Perform Matrix Multiplication

Enter the Matrices as B and A

- Enter each matrix, leaving at least one column between them.
- Highlight the cells that will contain the product matrix. This example result will be a 3 x 2 matrix. I selected cells A5:B7.
- Go to **fn** in the menu bar. The equal sign is automatically put into cell A5. Select **Math & Trig** on the left, **MMULT** on the right.
- Click **OK.**
- A box comes up asking where it should find the arrays you wish to use for the multiplication. You can click in **Array 1**, then highlight the cells that contain matrix B. Click in **Array 2**, then highlight the cells that contain matrix A.

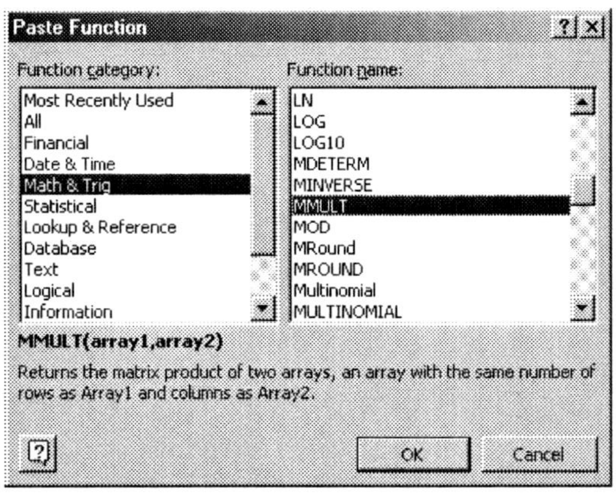

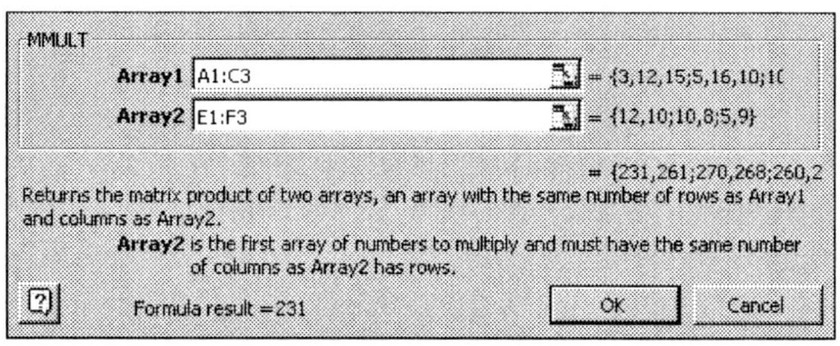

- Press **CNTRL SHIFT ENTER** on the keyboard to see the resulting matrix. ***DO NOT CLICK OK.***
- The resulting screen is shown below.
- Note that if you click OK, you will only see the result in the first cell, 231.

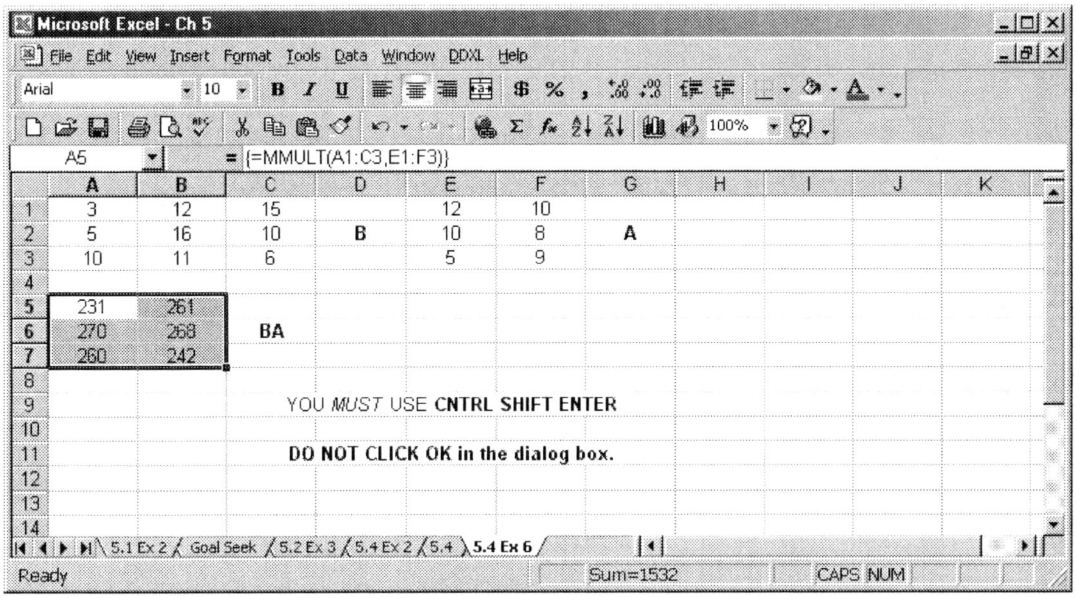

5.4 Example 3 on the TI-83 and TI-83 PLUS

This example asks you to find the inverse of matrix $A = \begin{bmatrix} -2 & 1 & 2 \\ 1 & 0 & -1 \\ 4 & -2 & -3 \end{bmatrix}$.

Find the Inverse of a Square Matrix

Enter the Matrix

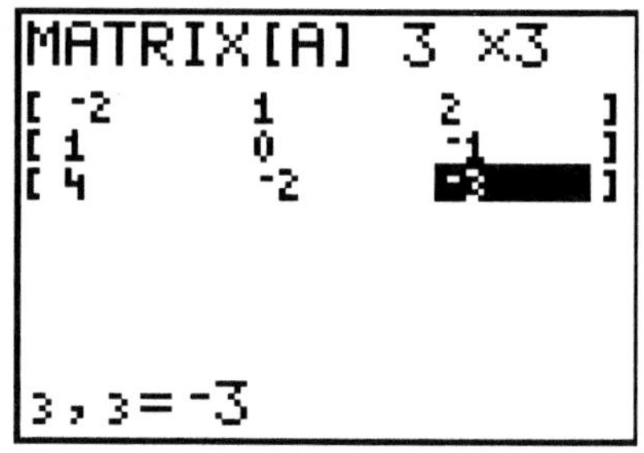

- Press [2nd] [x^{-1}] to access the MATRIX EDIT menu.
- Move the cursor to 1: [A] and press [ENTER].
- Enter the dimension of matrix A as 3 x 3. Enter the values into the matrix as shown.
- Note that the position is given at the bottom of the screen as 1,1= - 2 etc.
- Press [2nd] [MODE] to QUIT and return to the Home screen.

Find the Matrix Inverse on the Home Screen

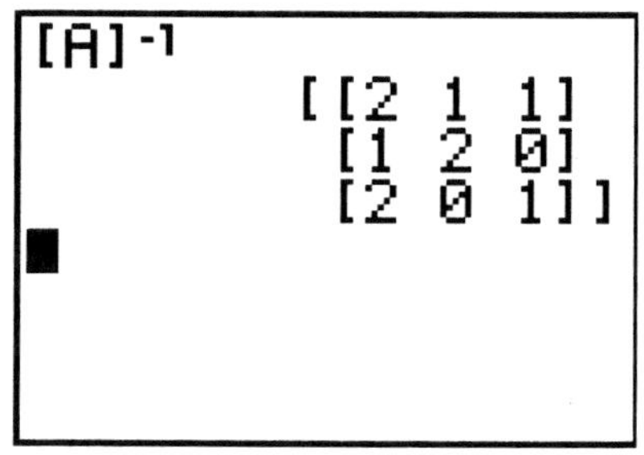

- Press [2nd] [MODE] to QUIT and return to the Home screen.
- Use MATRIX NAMES to enter the name of the matrix.
- Press the [x^{-1}] key to find the inverse of matrix A.
- Press [ENTER] to see A^{-1}.

Show that $AA^{-1} = I$

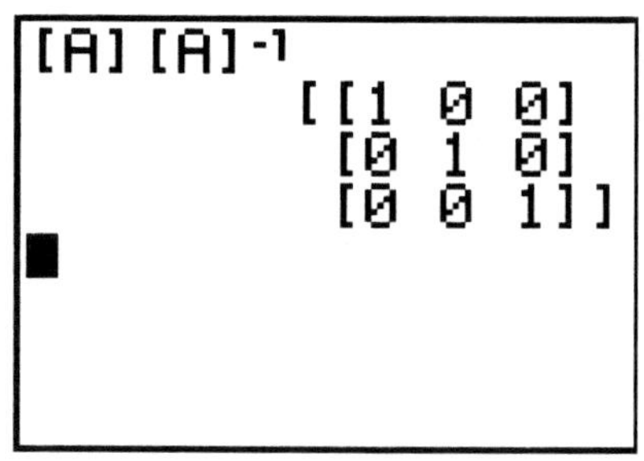

- On the Home screen use MATRIX NAMES to enter matrix A.
- On the same line, use MATRIX NAMES to enter matrix A again, then press the [x^{-1}] key.
- Press [ENTER] to see to see the product matrix, the identity matrix for a 3x3 matrix.

5.4 Example 3 on the TI-86

Example 5 asks you to find the inverse of matrix $A = \begin{bmatrix} -2 & 1 & 2 \\ 1 & 0 & -1 \\ 4 & -2 & -3 \end{bmatrix}$.

Find the Inverse of a Square Matrix

Enter the Matrix

- Press [2nd] [7] to access the MATRIX menu. Press [F2] to go to EDIT.
- You are asked to give the matrix a name. If you already used A for the name of a matrix, press the f key below it. Otherwise, use the [LOG] key to name it as matrix A.
- Press [ENTER]. Enter the dimension of matrix A as 3 x 3. Enter the values into the matrix as shown. Note that the position is given at the bottom of the screen.
- Press [EXIT] to return to the Home screen.

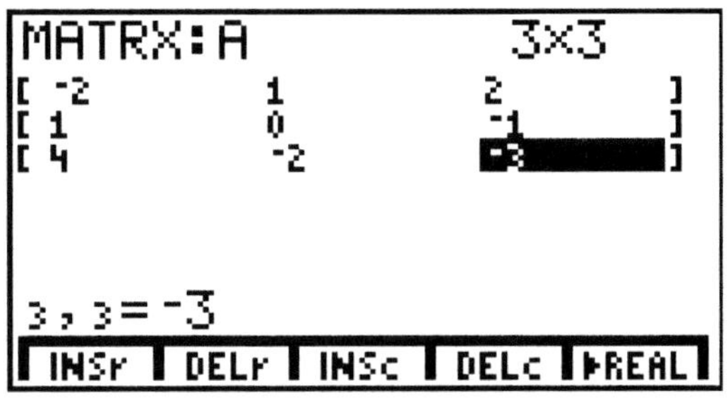

Find the Matrix Inverse on the Home Screen

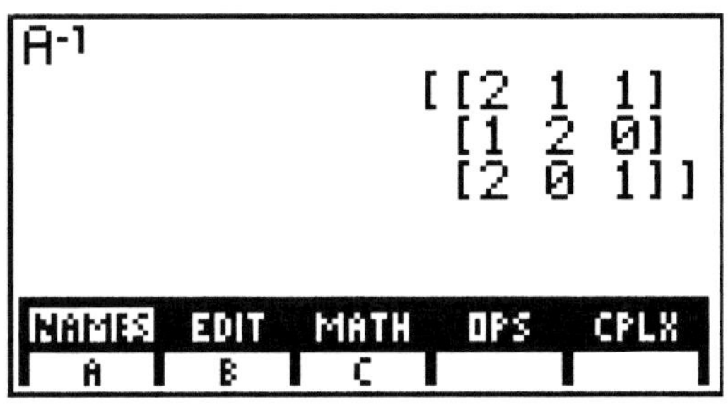

- Press [EXIT] to return to the Home screen.
- Use MATRIX NAMES to enter the name of matrix A. You may also use [ALPHA] [LOG] to enter the name of the matrix as A, etc.
- Press [2nd] [EE] to use the inverse function x^{-1}.
- Press [ENTER] to see A inverse.

Show that $AA^{-1} = I$

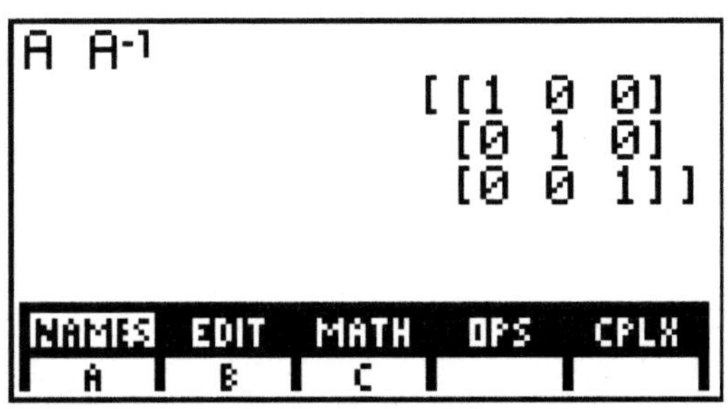

- On the Home screen use MATRIX NAMES to enter matrix A.
- On the same line, use MATRIX NAMES to enter matrix A again, then press the [x^{-1}] key.
- Press [ENTER] to see to see the product matrix, the identity matrix for a 3x3 matrix.

5.4 Example 3 using EXCEL

Example 5 asks you to find the inverse of matrix $A = \begin{bmatrix} -2 & 1 & 2 \\ 1 & 0 & -1 \\ 4 & -2 & -3 \end{bmatrix}$.

Find the Inverse of a Square Matrix

Enter the Matrix as A

- Enter the matrix.
- Highlight the cells that will contain the inverse matrix. This example result will be a 3 x 3 matrix. I selected cells A5:C7.
- Go to **fn** in the menu bar. The equal sign is automatically put into cell A5. Select **Math & Trig** on the left, **MINVERSE** on the right.
- Click **OK.**
- A box comes up asking where it should find the array for which you wish find the inverse. You can click in **Array** , then highlight the cells that contain matrix A.

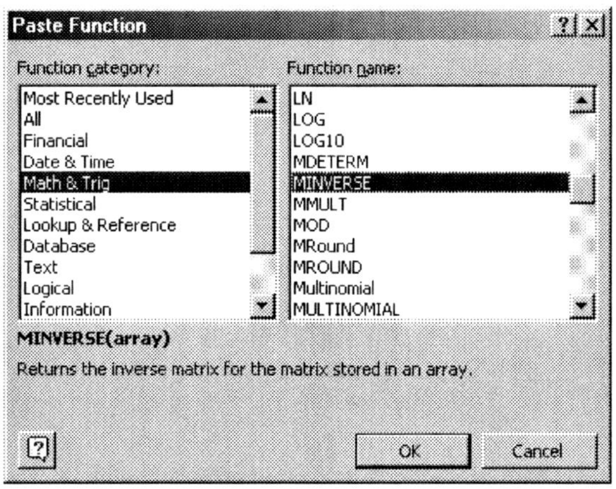

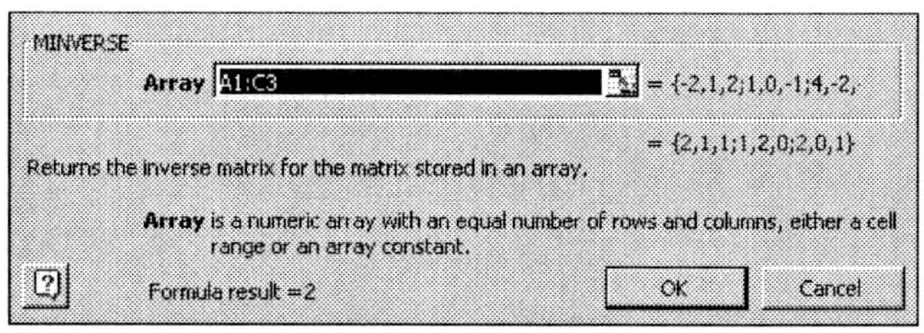

- Press **CNTRL SHIFT ENTER** on the keyboard to see the resulting matrix. ***DO NOT CLICK OK.***
- The resulting screen is shown below.
- Note that if you click OK, you will only see the result in the first cell 2.

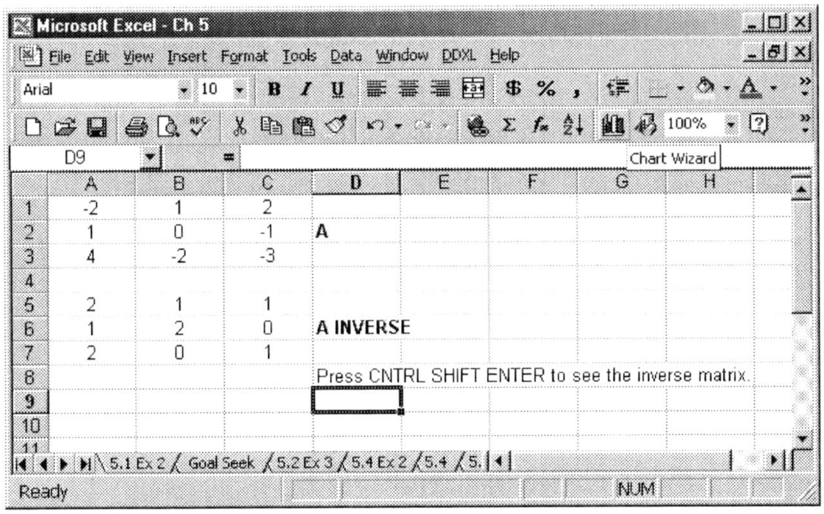

5.4 Example 8 on the TI-83, TI-83 PLUS, and TI-84 PLUS

This example asks you to solve the matrix equation $\begin{bmatrix} 5 & 7 & 12 \\ 2 & 3 & 5 \\ 1 & 1 & 1 \end{bmatrix}\begin{bmatrix} x \\ y \\ z \end{bmatrix} = \begin{bmatrix} 134{,}000 \\ 56{,}000 \\ 14{,}000 \end{bmatrix}$.

Name the coefficient matrix A, and the constant matrix C.

Solve a Matrix Equation

Enter the Matrices A and C

- Press [2nd] [x^{-1}] to access the MATRIX EDIT menu.
- Enter matrix A.
- Repeat the process and enter matrix C.
- Press [2nd] [MODE] to QUIT and return to the Home screen.

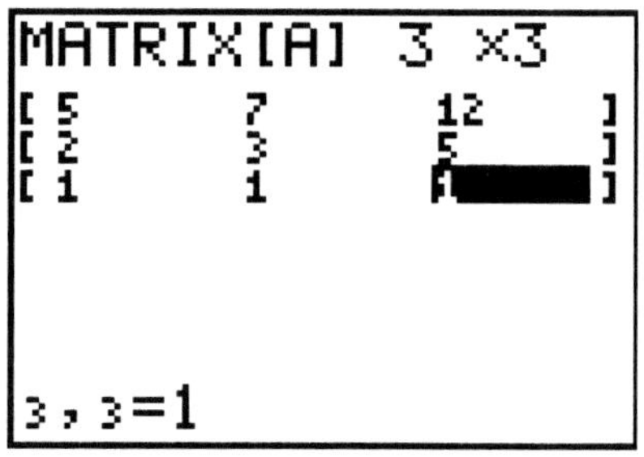

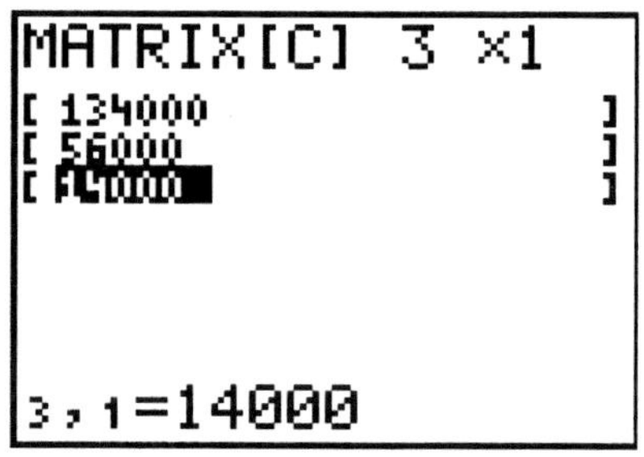

Find the product matrix of A^{-1} and C on the Home Screen

- Press [2nd] [MODE] to QUIT and return to the Home screen.
- Use MATRIX NAMES to enter the name of the matrix A.
- Use the [x^{-1}] key to place the inverse sign after of matrix A as shown. Use MATRIX NAMES to enter the name of the matrix C. Note that you do not need to see the inverse matrix of A.
- Press [ENTER] to see the product.

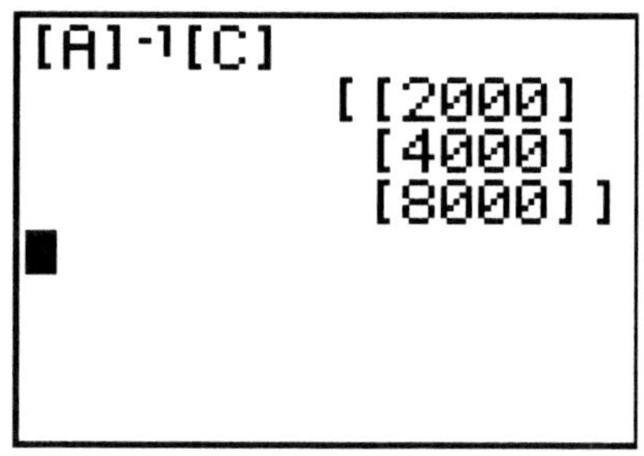

5.4 Example 8 on the TI-86

This example asks you to solve the matrix equation $\begin{bmatrix} 5 & 7 & 12 \\ 2 & 3 & 5 \\ 1 & 1 & 1 \end{bmatrix}\begin{bmatrix} x \\ y \\ z \end{bmatrix} = \begin{bmatrix} 134,000 \\ 56,000 \\ 14,000 \end{bmatrix}$.

Name the coefficient matrix A, and the constant matrix C.

Solve a Matrix Equation

Enter the Matrices A and C

- Press [2nd] [7] to access the MATRIX menu. Press [F2] to go to EDIT.
- Enter matrix A.
- Repeat the process and enter matrix C.
- Press [EXIT] to return to the Home screen.

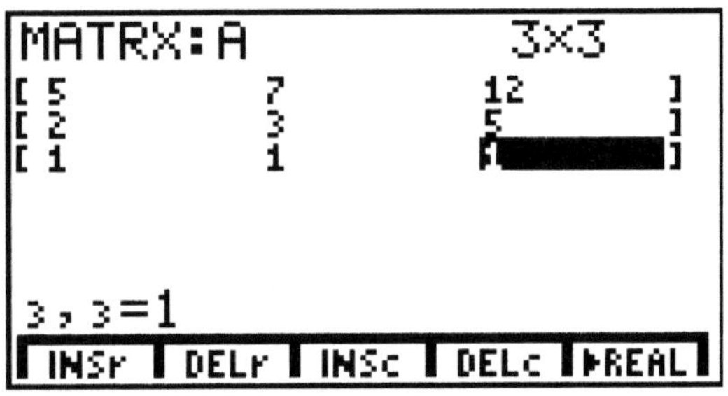

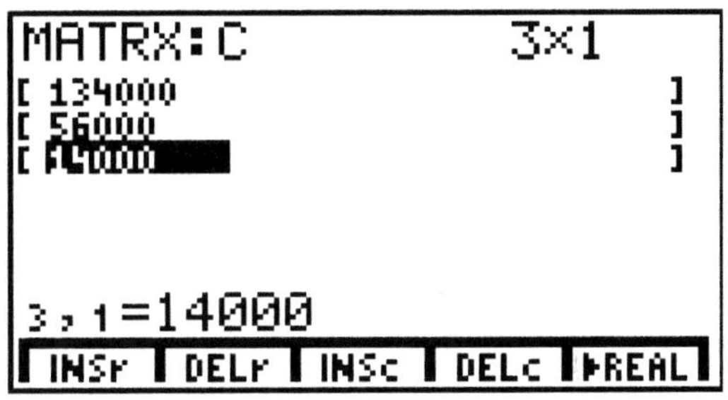

Find the product matrix of A^{-1} and C on the Home Screen

- Press [EXIT] to return to the Home screen.
- Use MATRIX NAMES or [ALPHA] [LOG] to enter the name of matrix A.
- Press [2nd] [EE] to use the inverse function x^{-1}.
- Use MATRIX NAMES or [ALPHA] [COS] to enter the name of matrix C. Note that you do not need to see the inverse matrix of A.
- Press [ENTER] to see the product.

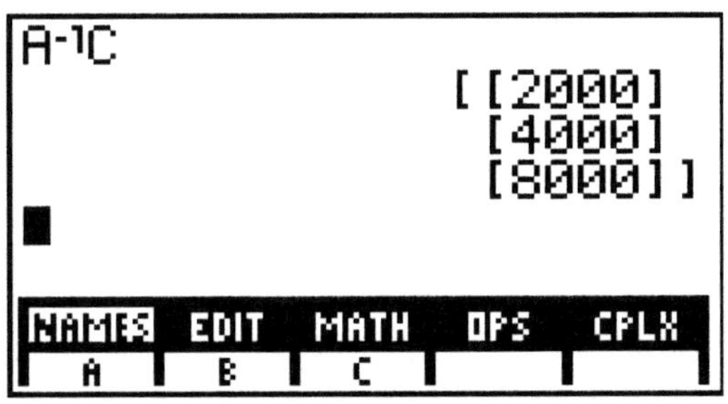

5.4 Example 8 using EXCEL

This example asks you to solve the matrix equation $\begin{bmatrix} 5 & 7 & 12 \\ 2 & 3 & 5 \\ 1 & 1 & 1 \end{bmatrix}\begin{bmatrix} x \\ y \\ z \end{bmatrix} = \begin{bmatrix} 134,000 \\ 56,000 \\ 14,000 \end{bmatrix}$.

Name the coefficient matrix A, and the constant matrix C.

Solve a Matrix Equation

Enter the Matrices A and C

- Enter the matrices.
- Highlight the cells that will contain the inverse matrix. This example result will be a 3 x 3 matrix. I selected cells A5:C7.

Find A Inverse

- Go to **fn** in the menu bar. The equal sign is automatically put into cell A5. Select **Math & Trig** on the left, **MINVERSE** on the right.
- Click **OK.**A box comes up asking where it should find the array for which you wish find the inverse. You can click in **Array** , then highlight the cells that contain matrix A.

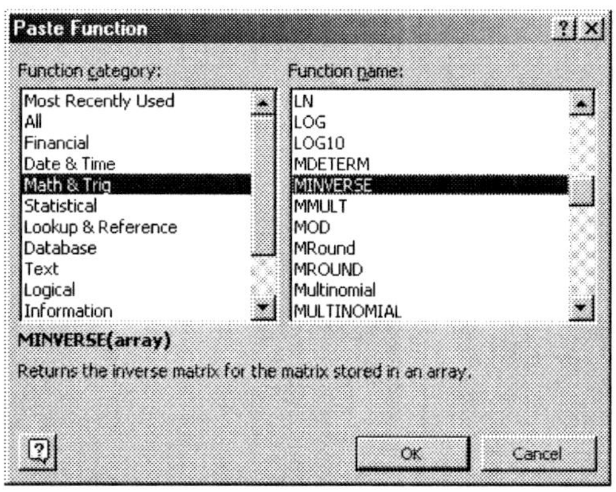

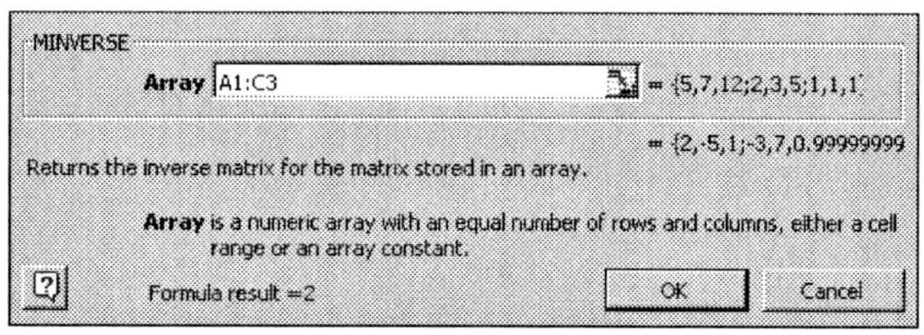

- Press **CNTRL SHIFT ENTER** on the keyboard to see the resulting matrix. ***DO NOT CLICK OK.***
- The resulting screen is shown below.
- Note that if you click OK, you will only see the result in the first cell 2.

Find the product of A Inverse and C

- Highlight the cells that will contain the product matrix. This example result will be a 3 x 1 matrix. I selected cells A9:A11.
- Go to **fn** in the menu bar. The equal sign is automatically put into cell A9. Select **Math & Trig** on the left, **MMULT** on the right.
- A box comes up asking where it should find the arrays you wish to use for the multiplication. You can click in **Array 1**, then highlight the cells that contain matrix A Inverse. Click in **Array 2**, then highlight the cells that contain matrix C.

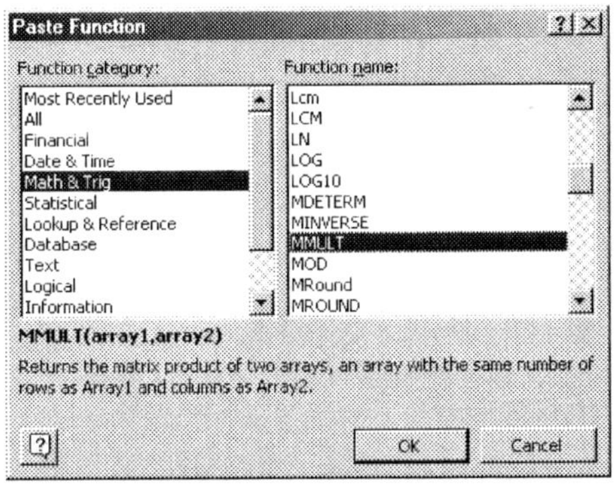

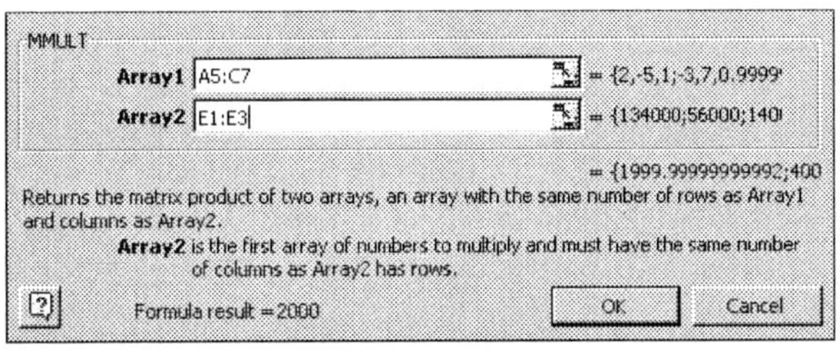

- Press **CNTRL SHIFT ENTER** on the keyboard to see the resulting matrix. ***DO NOT CLICK OK.***
- The resulting screen is shown below.
- Note that if you click OK, you will only see the result in the first cell, 2000.

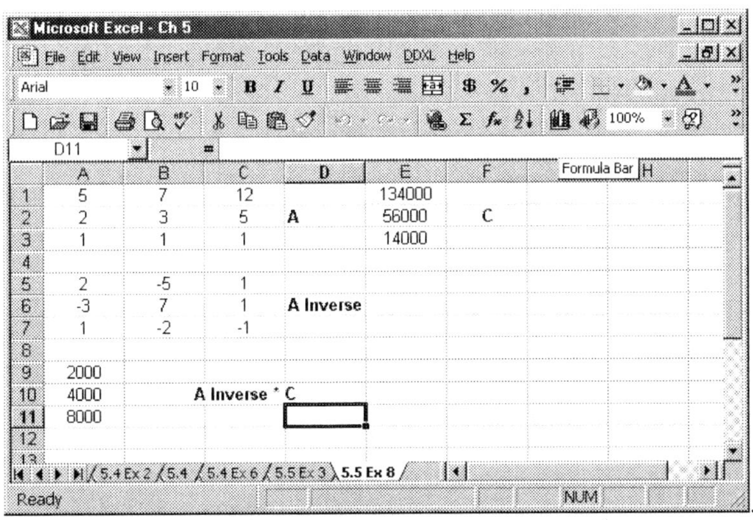

CHAPTER 6

SPECIAL TOPICS

6.1 Example 1 on the TI-83, TI-83 Plus, and TI-84 PLUS

This example asks you to graph the solution of the inequality $6x-3y\le 15$. To do this, first solve the inequality for y to obtain $y\ge 2x-5$. Graph $y=2x-5$. Determine whether or not the solution includes the line, and then format the graph as either connected with line segments or dotted. Finally, take a test point and use the Shade feature of the calculator to shade the solution region.

Solve the linear inequality $6x-3y\le 15$ by graphing.

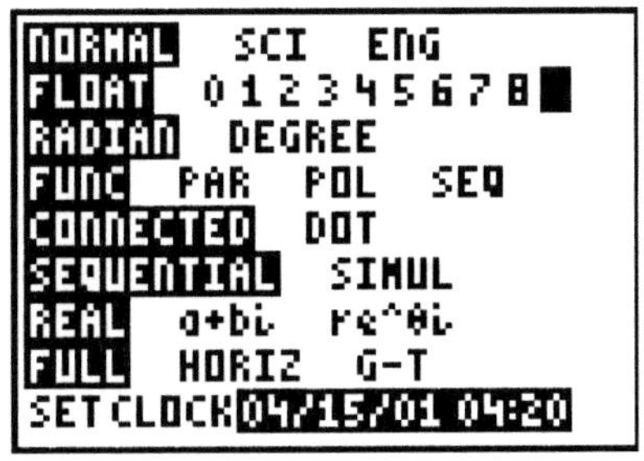

Select solid or dotted line using MODE

- Since this is a ≤ inequality, the graph is a solid line. To graph a solid line, press the [MODE] key. Cursor down to the line that reads Connected Dot. Select Connected.
- Press [CLEAR] to return to the Home screen or [Y=] to go to the Y= menu.

Enter the function

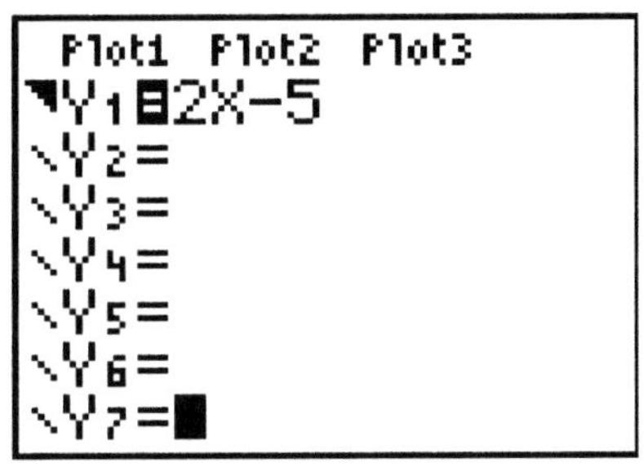

- Turn off or clear any equations or plots.
- Solve the inequality for y to get $y\ge 2x-5$.
- Enter Y1= $2x-5$.
- Set the window to [-10, 10, 1] by [-10, 10, 1].

Shade the solution region

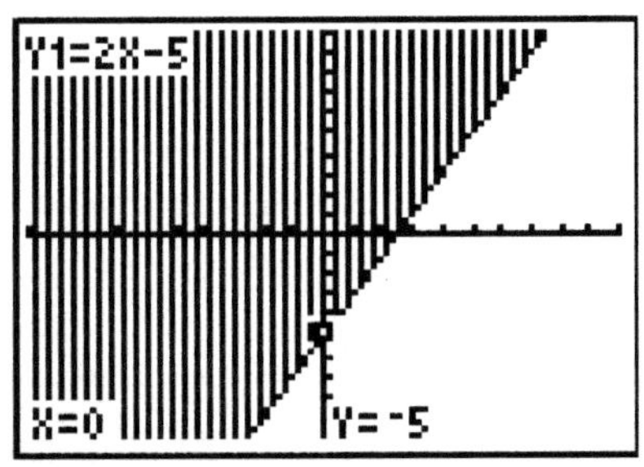

- Use a test point such as $(0,0)$ to decide whether the solution is above or below the line. This solution is above the line.
- Move the cursor to the very left of the ⋱Y= line. The ⋱ will be flashing.
- Press [ENTER] until you see ◥ symbol. This means that you want the graph to be shaded above the line. Move the cursor away from the symbol to select it.
- Press [GRAPH] to see the shaded graph.

6.1 Example 1 on the TI-86

This example asks you to graph the solution of the inequality $6x-3y\le 15$. To do this, first solve the inequality for y to obtain $y\ge 2x-5$. Graph $y=2x-5$. Determine whether or not the solution includes the line, and then format the graph as either connected with line segments or dotted. Finally, take a test point and use the Shade feature of the calculator to shade the solution region.

Solve the linear inequality $6x-3y\le 15$ by graphing.

Select solid or dotted line using MODE

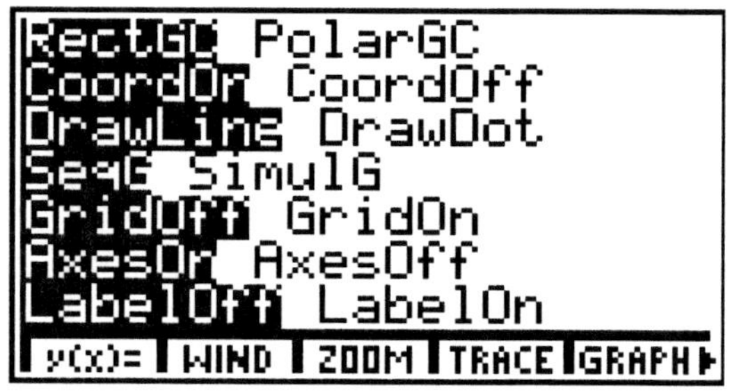

- Since this is a $\le$ inequality, the graph is a solid line. To graph a solid line, press the [GRAPH] key if you are not in the graph menu.
- Press [MORE] [F3] to select FORMT.
- Cursor down to the line that reads DrawLine DrawDot.
- Select DrawLine.
- Press [CLEAR] to return to the Home screen or [MORE] until you can select the y= menu.

Enter the function

- Turn off or clear any equations or plots.
- Solve the inequality for y to get $y\ge 2x-5$.
- Enter y1= $2x-5$.
- Set the window to [-10, 10, 1] by [-10, 10, 1].

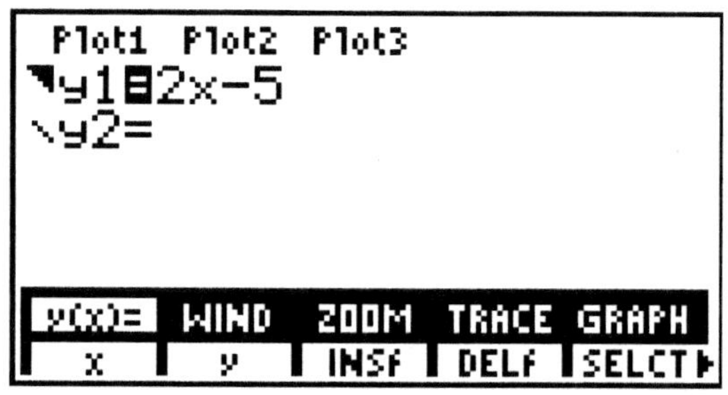

Shade the solution region

- Use a test point such as $(0,0)$ to decide whether the solution is above or below the line. This solution is above the line.
- In the y= menu press [MORE] then [F3] to select STYLE.
- Press [F3] STYLE until you see ◥ symbol. This means that you want the graph to be shaded above the line.
- Press [EXIT] [F5] to see the shaded graph.

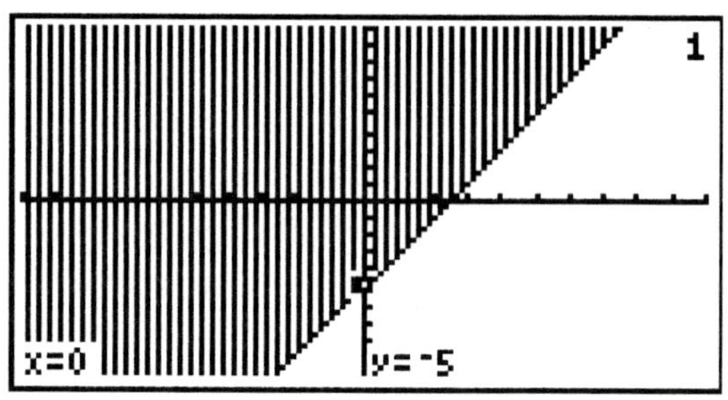

6.1 Example 5 on the TI-83 and TI-83 Plus

This example asks you to find and graph the solution of the 4 inequalities and find all the corners of the resulting solution region.

$$y \le 71 - x$$
$$y \le \frac{1260 - 15x}{28}$$
$$x \ge 0$$
$$y \ge 0$$

Graph the first two inequalities as Y1 and Y2. The last two inequalities $x \ge 0$, $y \ge 0$ can be handled by using a window that has Xmin = 0 and Ymin = 0.

Determine whether or not the solutions include the line, and then format the graph as either connected with line segments or dotted. Since both inequalities include the line, use line segments. Finally, take a test point and use the Shade feature of the calculator to shade the solution region.

Solve the system of linear inequalities $\begin{array}{l} y \le 71 - x \\ y \le \dfrac{1260 - 15x}{28} \\ x \ge 0 \\ y \ge 0 \end{array}$ **by graphing.**

Select solid or dotted line using MODE

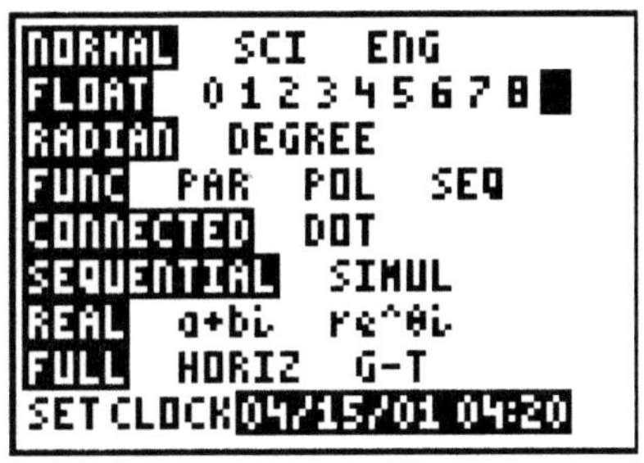

- Since both inequalities include the line, graph each line as solid. Press the [MODE] key, select Connected.
- Press [CLEAR] to return to the Home screen or [Y=] to go to the Y= menu.

Enter the functions

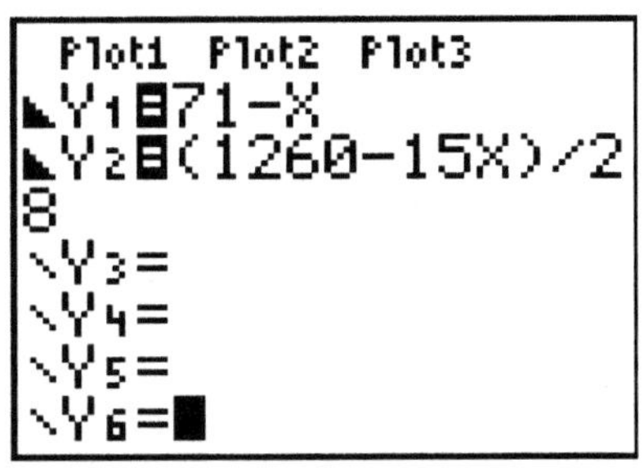

- Turn off or clear any equations or plots.
- Solve the equation for y and enter it as Y1. Enter Y1=$71 - x$.
- Enter Y2=$(1260 - 15x)/28$.
- Set the window to [0, 100, 10] by [0, 100, 10].

Shade the solution region

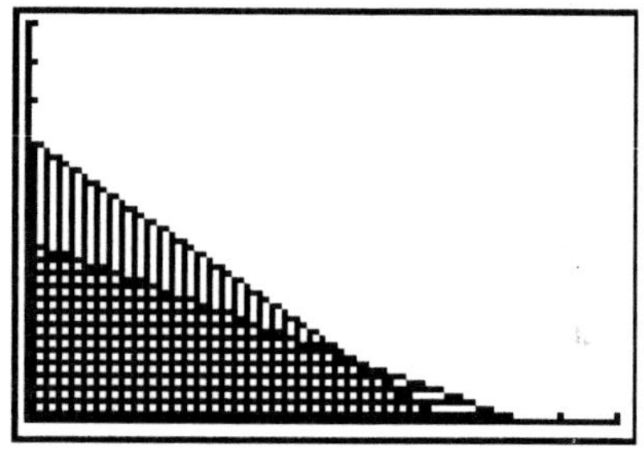

- Use a test point for each inequality to decide whether the solution is above or below the line. Both solutions are below the line. Move the cursor to the very left of the ⋱Y1= line. The ⋱ will be flashing.
- To select the shading below the line for Y1, press [ENTER] until you see the ◣ symbol. This means that you want the graph to be shaded below the line. Move the cursor away from the symbol to select it.
- To select the shading above the line for Y2, press [ENTER] until you see the ◣ symbol. This means that you want the graph to be shaded below the line. Move the cursor away from the symbol to select it.
- Press [GRAPH] to see the shaded graphs. The solution to the system of inequalities is the intersection of the two shaded areas, which is the darker trapezoidal shape.

Find the corner points

- There are three corner points: the intersection of Y1 and Y2, the zero of Y1, and the y-intercept of Y2.
- Use [2nd] [TRACE] CALC 5:intersect to find the intersection of Y1 and Y2 at $(56,15)$.
- Use the up arrow to Y1. Use [2nd] [TRACE] CALC 2:zero to find the zero of Y1 at $(71,0)$.
- Press the [TRACE] key.
- Enter the value 0 and press [ENTER].
- Press the up key until you are on Y2. The point $(0,45)$ is at the bottom of the screen. This is the third corner.

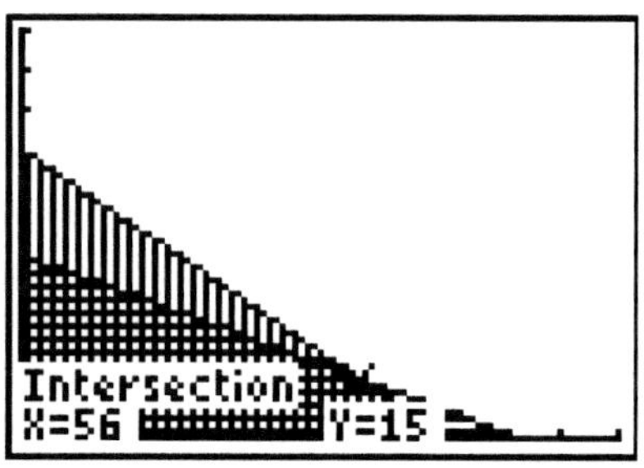

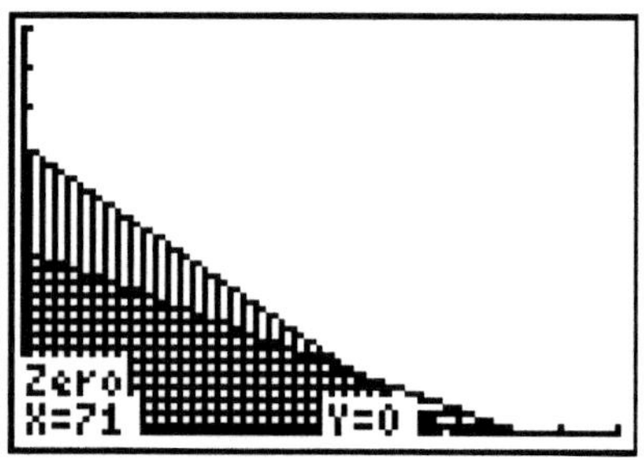

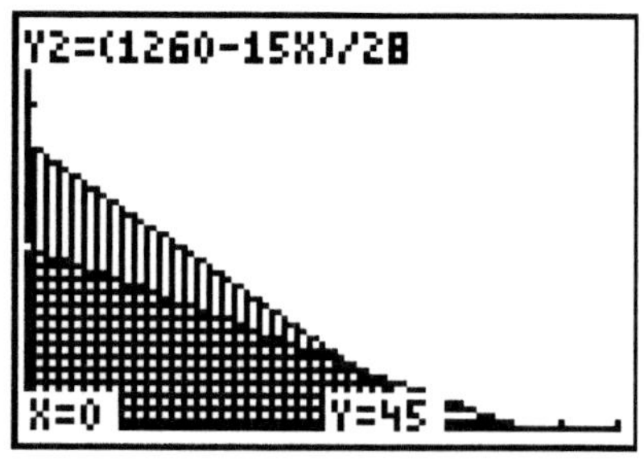

6.1 Example 5 on the TI-86

This example asks you to find and graph the solution of the 4 inequalities and find all the corners of the resulting solution region.

Graph the first two inequalities as y1 and y2. The last two inequalities $x \geq 0, \ y \geq 0$ can be handled by using a window that has xmin = 0 and ymin = 0.

Determine whether or not the solutions include the line, and then format the graph as either connected with line segments or dotted. Since both inequalities include the line, use line segments. Finally, take a test point and use the Shade feature of the calculator to shade the solution region.

Solve the system of linear inequalities

$$\begin{aligned} y &\leq 71 - x \\ y &\leq \frac{1260 - 15x}{28} \\ x &\geq 0 \\ y &\geq 0 \end{aligned}$$

by graphing.

Select solid or dotted line using MODE

- Since both inequalities include the line, graph each line as solid. To do this, press the [GRAPH] key if you are not in the graph menu.
- Press [MORE] [F3] to select FORMT.
- Cursor down to the line that reads DrawLine DrawDot.
- Select DrawLine.
- Press [CLEAR] to return to the Home screen or [MORE] until you can select the y = menu.

- Turn off or clear any equations or plots.
- Solve the equation for y and enter it as y1. Enter y1= $71 - x$.
- Enter y2= $(1260 - 15x)/28$.
- Set the window to [0, 100, 10] by [0, 100, 10].

Shade the solution region

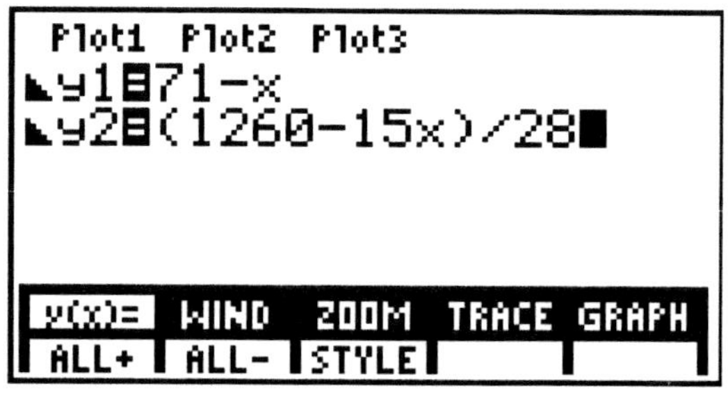

- Use a test point for each inequality to decide whether the solution is above or below the line. Both solutions are below the line.
- To shade below the line, in the y= menu press [MORE] then [F3] to select STYLE.
- Position the cursor on y1. Press [F3] STYLE until you see ◣ symbol. This means that you want the graph to be shaded below the line.
- Position the cursor on y2. Press [F3] STYLE until you see ◣ symbol. This means that you want the graph to be shaded below the line.
- Press [EXIT] [F5] to see the shaded graphs. The solution to the system of inequalities is the intersection of the two shaded areas. Press [CLEAR] to remove the menu line. Press [EXIT] to go back to the menu line.

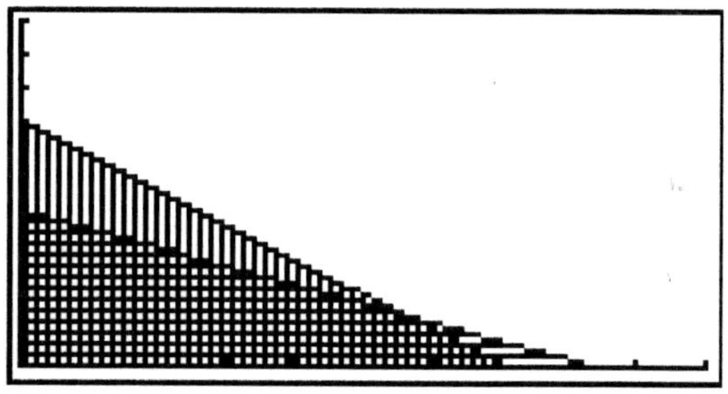

Find the corner points

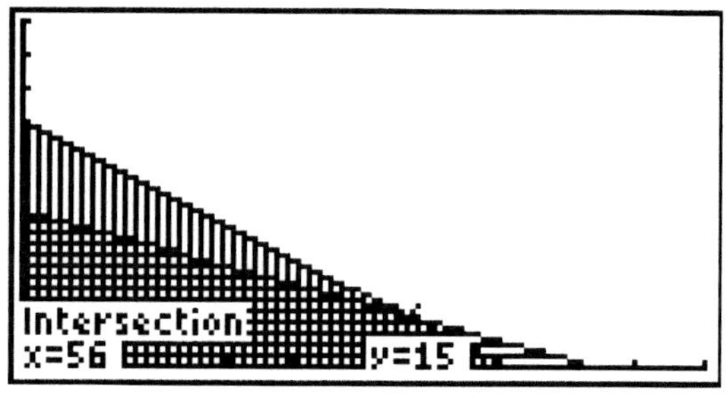

- There are three corner points: the intersection of Y1 and Y2, the zero of Y1, and the y-intercept of Y2.
- Press [MORE] [F1] MATH [MORE] [F3] ISECT to find the intersection of y1 and y2 at $(56,15)$.
- Press [EXIT] [F1] MATH [F1] ROOT to find the zero of y1 at $(71,0)$.
- Press [EXIT] [MORE] [F1] EVAL. Enter 0 at the bottom of the screen. Press [ENTER] to see the y-intercept of one of the functions. Move the up arrow until you are on y2 at $(0,45)$.

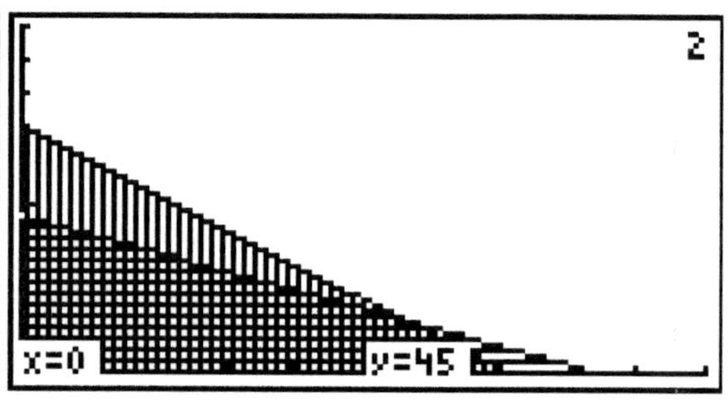

6.1 Example 5 on the TI-84 PLUS

Graph a System Linear Inequalities on the TI-84 Plus Silver using Inequalz Apps

This example asks you to find and graph the solution of the 4 inequalities and find all the corners of the resulting solution region. Enter the first two inequalities as Y1 and Y2. Enter Y3 ≥ 0 and X1 ≥ 0 .

Determine whether or not the solutions include the line, and then format the graph as either connected with line segments or dotted. Since both inequalities include the line, use line segments. Identify the points of intersection of the solution set.

Solve the system of linear inequalities
$$\begin{aligned} &y \leq 71 - x \\ &y \leq \frac{1260 - 15x}{28} \\ &x \geq 0 \\ &y \geq 0 \end{aligned}$$
by graphing.

- Press the APPS key. It is in the third row of keys next to the [MATH] key. Press the green [ALPHA] key, then the [x^2] key to jump to the I's.
- Select Inequalz.
- Enter Y1 $\leq 71 - x$ and Y2 $\leq (1260 - 15x)/28$, Y3 ≥ 0, and X1 ≥ 0.
- To enter the inequality symbols for Y1 and Y2, place the cursor over the = sign. A menu line with symbols appears at the bottom of the screen.
- Press the green [ALPHA] key and the key directly under the inequality you want.
- To select $\leq$, press [ALPHA] [ZOOM].
- Note that the inequality symbols are only shown when the cursor is over the = sign.
- Move the cursor over X= at the top of the screen and press [ENTER]. Enter X1 ≥ 0 .

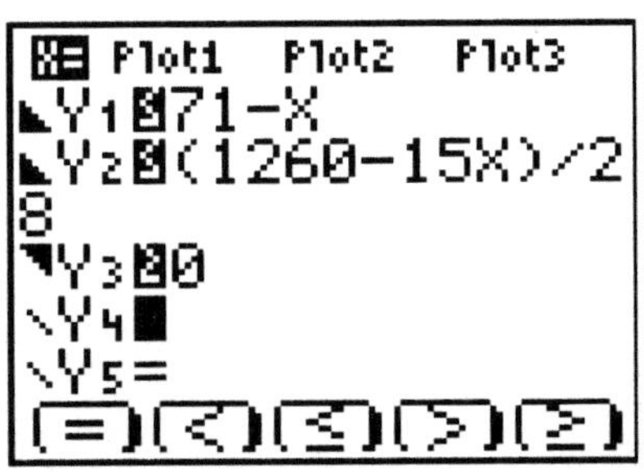

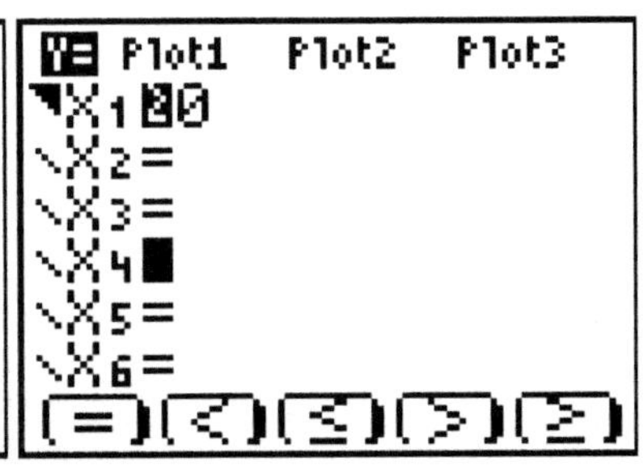

- Press [GRAPH] to see the graphs of the four inequalities.
- Press the green [ALPHA] key, then either the [Y=] or [WINDOW] key to select SHADES.
- Move the cursor up to 1: and press [ENTER] to select 1:Ineq Intersection.

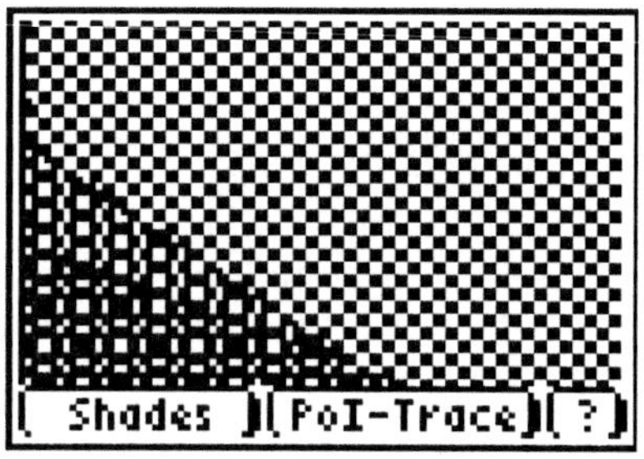

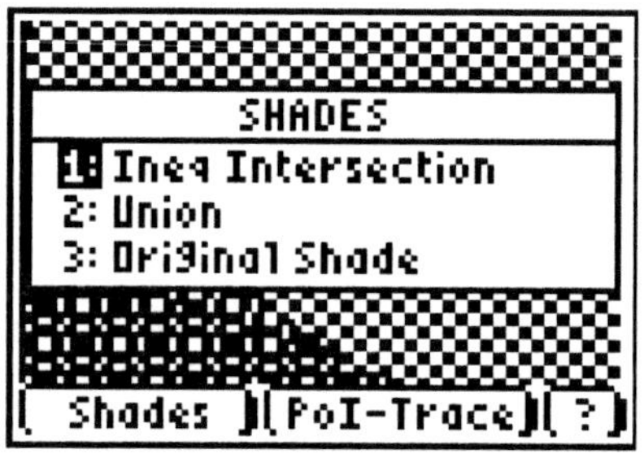

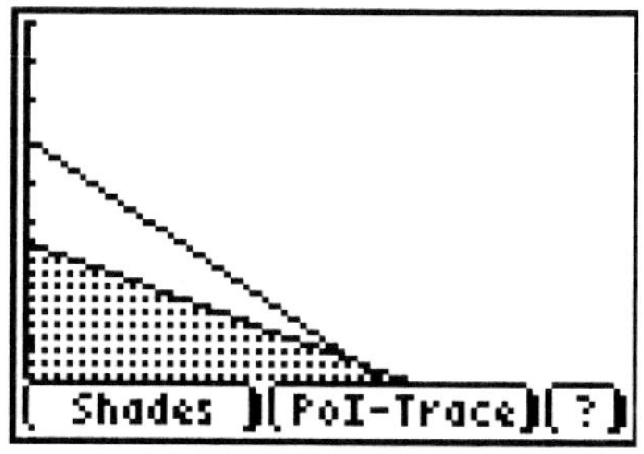

- Press green [ALPHA] key.
- Press either the [ZOOM] or [TRACE] key to use PoI-Trace to locate any intersection points of the linear system. Note that you will get the coordinates of all possible combinations of intersections of the four inequalities, two of which, $(0,71)$ and $(84,0)$ are not solutions of the system.

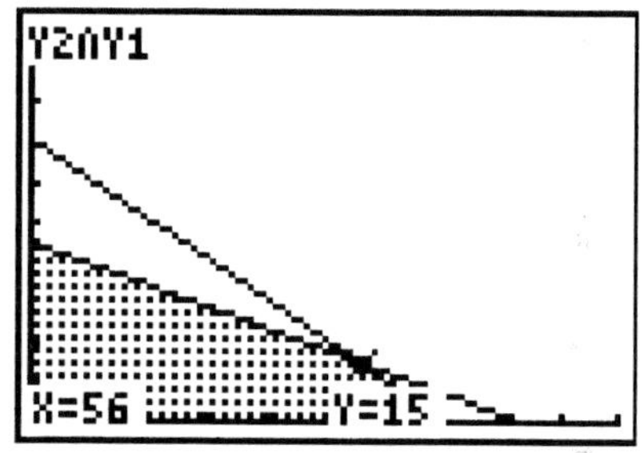

- Use the arrow keys to locate the four corner points of the solution, $(0,0)$, $(0,45)$, $(56,15)$, $(71,0)$.
- Note that the two inequalities that are intersected at shown by their names at the top of the screen.

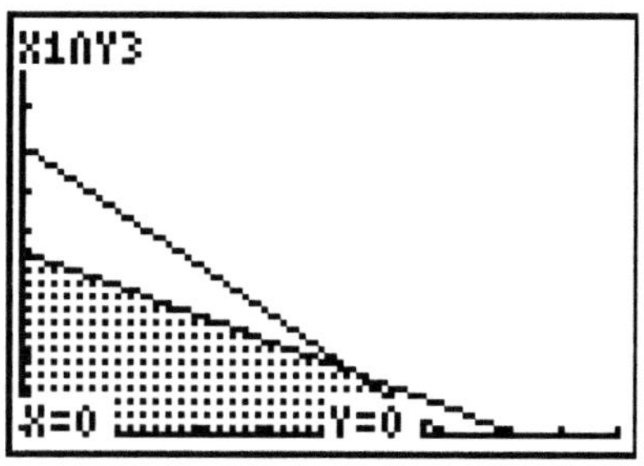

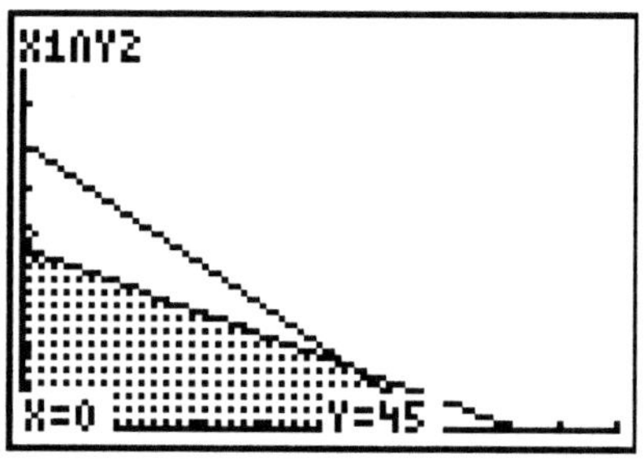

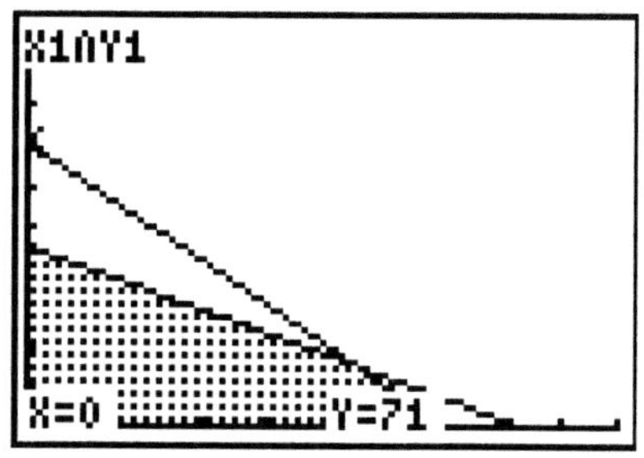

To Quit Inequalz

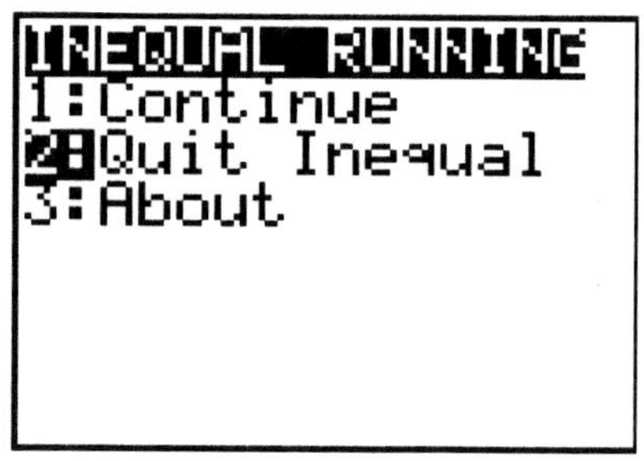

- To quit Inequalz, press APPS [2nd] [x^2] to jump to the I's.
- Press [ENTER].
- Move the cursor to 2: Quit Inequalz and press [ENTER] .

6.3 EXAMPLE 1 on the TI-83, TI-83 PLUS, and TI-84 PLUS

This example asks you to find the first four terms of two sequences,

a. $a_n = \frac{8}{n}$ b. $b_n = (-1)^n n(n+1)$.

We will first put the calculator in sequence mode, calculate the first four terms of each sequence, and examine graph and table of each sequence.

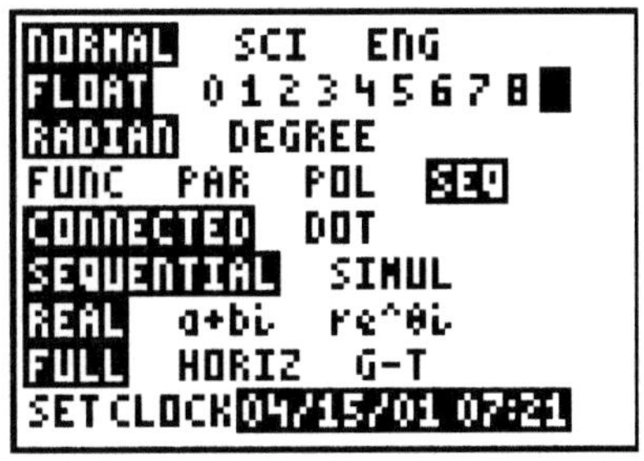

Sequence Mode

- To put the calculator into sequence mode, press the [MODE] key.
- On the 4th line select Seq.

Calculate the Terms of the Sequence

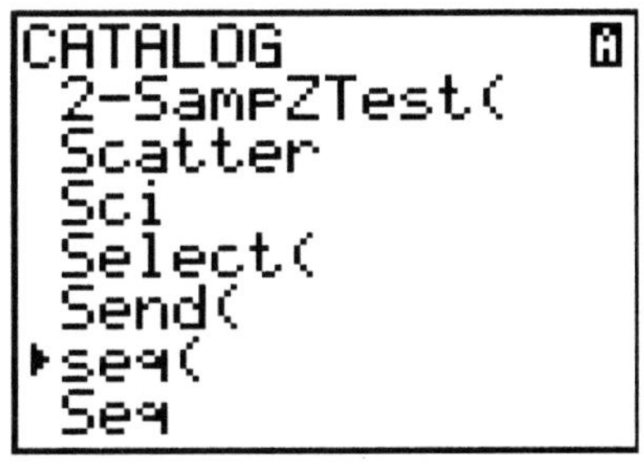

- Press [CLEAR] to return to the Home screen.
- One way to access the seq(function is to press [2nd] [0] to access the CATALOG.
- Press the [LN] key to jump to the S's. Cursor down until you find seq(. Press [ENTER] to puts seq(on the Home screen.

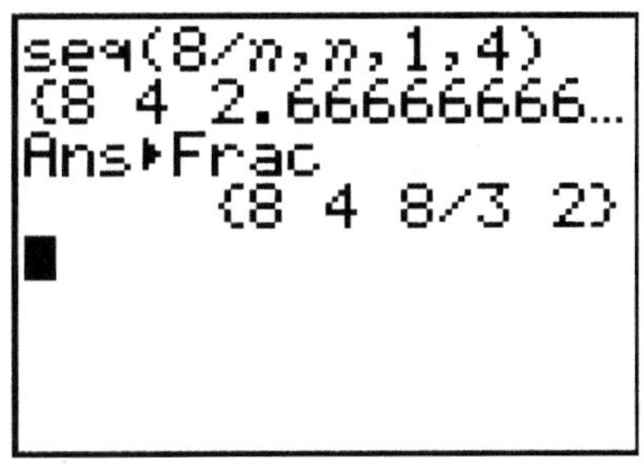

- Another way to access the seq(function is to press [2nd] [STAT] and go to LIST OPS. Select 5:seq(. This puts seq(on the Home screen.
- Enter the expression for each sequence, the variable you are using, the beginning and end of the sequence, and finally the increment if needed. Use the [X,T,Θ,n] key to enter the variable *n*.
- Press [ENTER] to see the sequence.
- To change the sequence to fraction form, press the [MATH] key, and in the MATH menu select 1:→Frac.
- The screens for each sequence are shown.

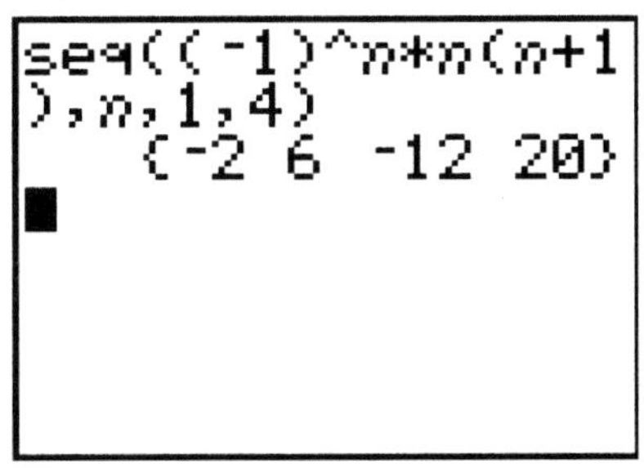

Graph the sequence $a_n = \frac{8}{n}$

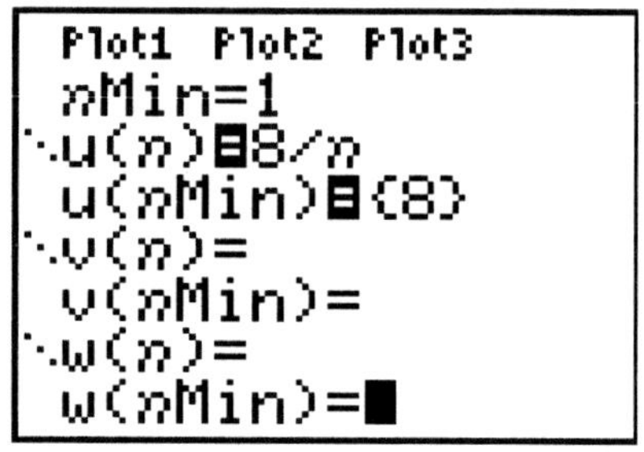

- The basics of graphing a sequence are the same as graphing a function, namely, you must enter the expression that defines the sequence, set the window, etc. The settings, however, look very different. Here is the screen for entering the first sequence, $a_n = \frac{8}{n}$. Note that the sequence functions are denoted as *u, v,* and *w*.

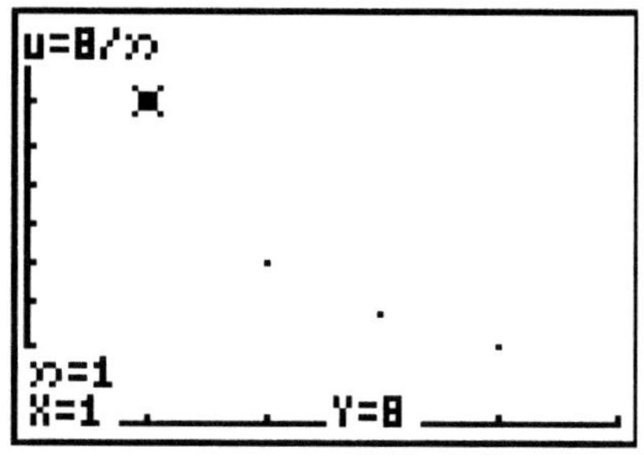

- You must enter a value for the lowest value of n you want to use as nMin. We use nMin = 1.
- Next, enter the expression for the sequence $u(n) = 8/n$. Use the [X,T,Θ,n] key to enter n.
- Finally, you must enter the first term of the sequence $u(nMin) = 8$
- The calculator will insert the braces when you press [ENTER]. $u(nMin) = \{8\}$.
- Press [GRAPH] to see the graph. Use [TRACE] to move from point to point.

Set the Window for the Graph of the Sequence

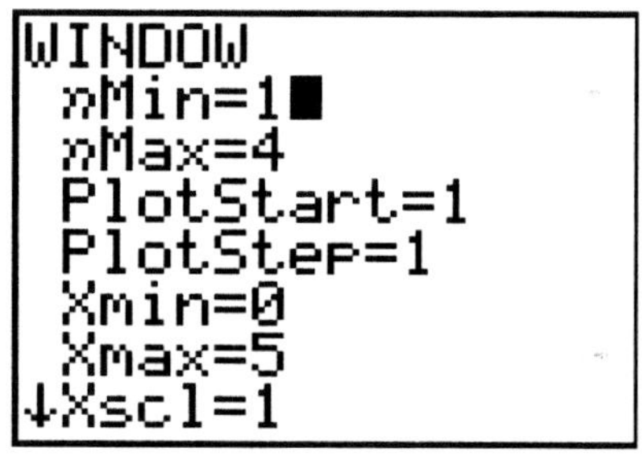

- Some of the window settings for a sequence are different from the window settings of functions.

  nMin = 1 and nMax = 4 since we want the first four terms of the sequence.

  PlotStart = 1 because we want to start at the first term.

  PlotStep = 1 because we want to see terms 1, 2, 3, and 4, an increment or step of 1 from term to term.

  The remaining selections are the same as for functions.
- The window settings are shown over two screens.

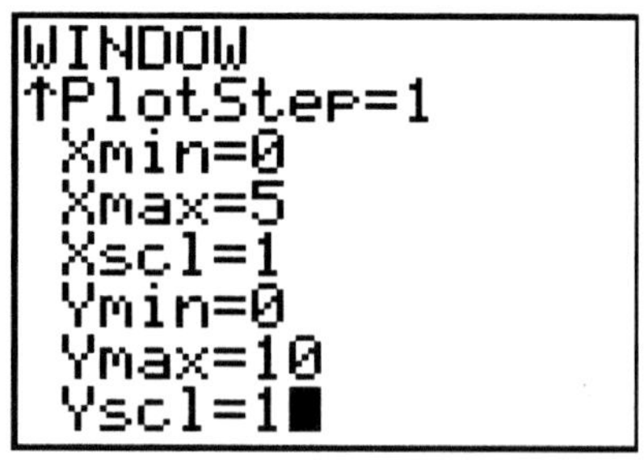

Create a Table of Values for the sequence

- Press [2nd] [WINDOW] to set up the table as TABLE ASK.
- Press [2nd] [GRAPH] to enter the values of n needed.
- Note that you may use TABLE AUTO, but you get more terms than you needed.

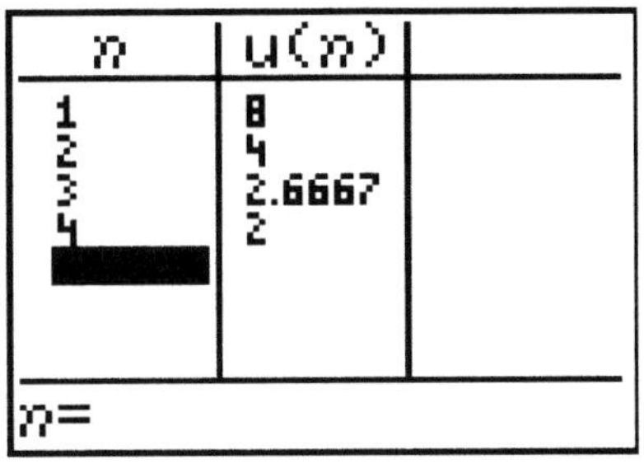

Create the graph and table for $b_n = (-1)^n n(n+1)$

- The only change in the window settings for this sequence are in the y-values of the function. We know that the y-values range from -12 to 20, so set Ymin = -15, Ymax = 25, and Yscl=5.
- Here are the window settings over 2 screens.

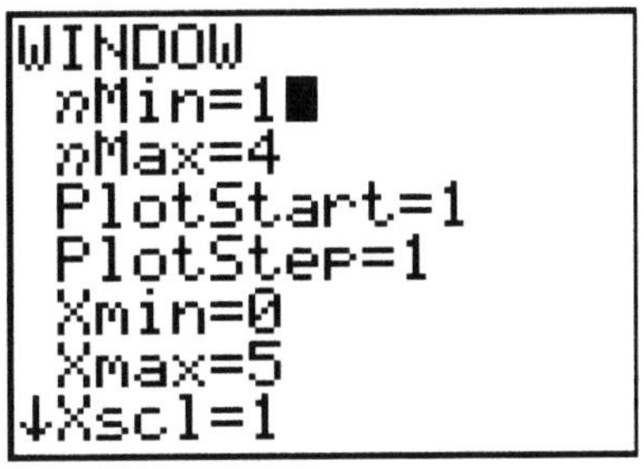

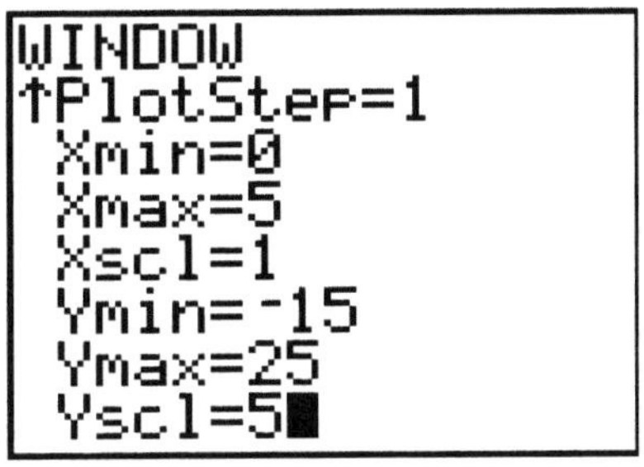

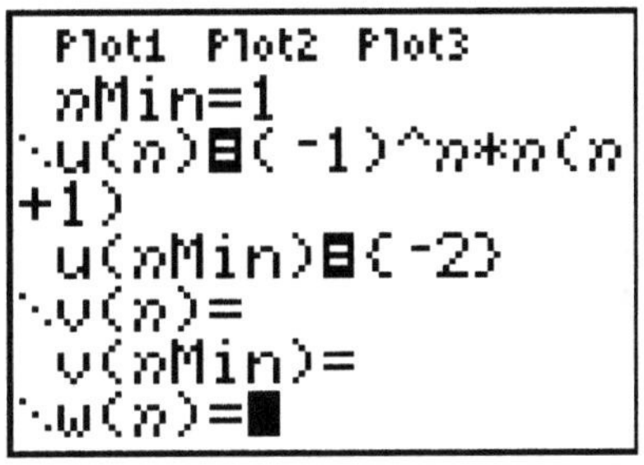

- To create the graph, turn off or clear $u(n)$. At $u(n)$ enter $(-1)^n * n(n+1)$. Use the [X,T,Θ,n] key to enter n.
- nMin is still set at 1, set $u(n\text{Min})$ = - 2, the value of the first term of the sequence.
- Create the table using TABLE ASK.

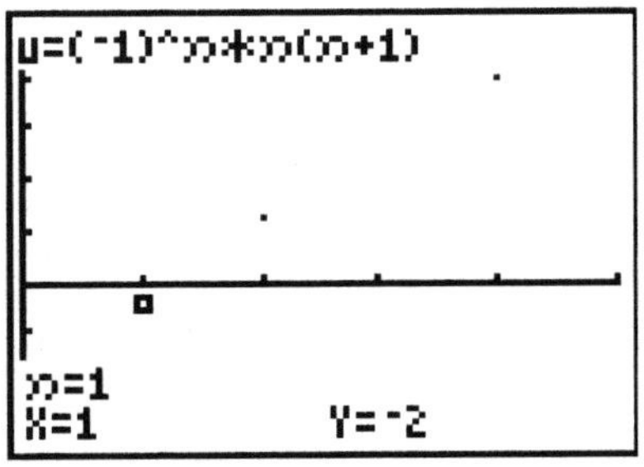

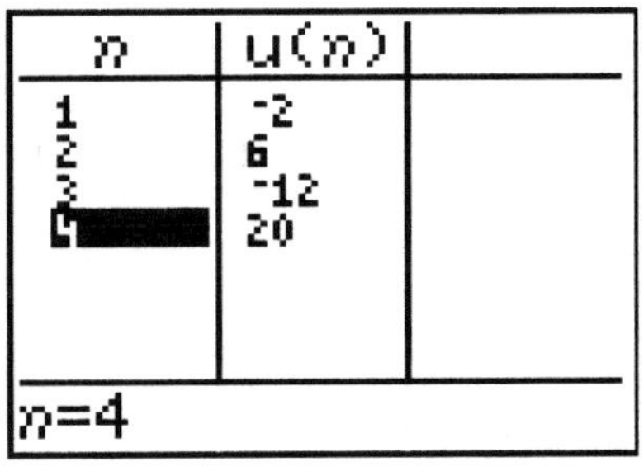

6.3 EXAMPLE 1 on the TI-86

This example asks you to find the first four terms of two sequences,

a. $a_n = \dfrac{8}{n}$ b. $b_n = (-1)^n n(n+1)$.

We will calculate the first four terms of each sequence, and examine graph of each sequence.

Sequence Mode

The TI-86 does not have a sequence mode. You can calculate the terms of a sequence on the Home screen. You can graph the sequence using one of the Plots, so enter the sequence x and y values into the lists L_1 and L_2.

Calculate the Terms of the Sequence

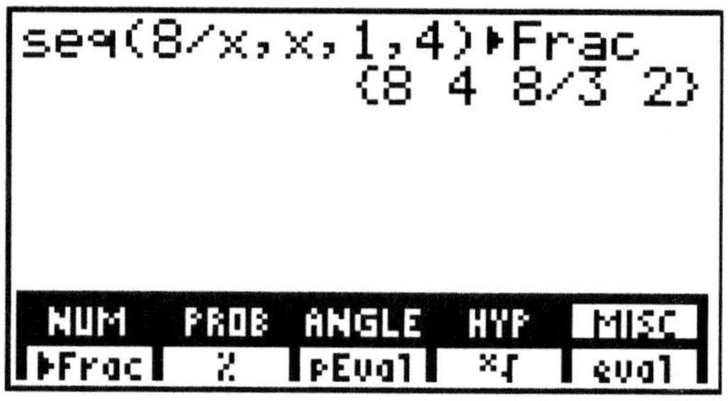

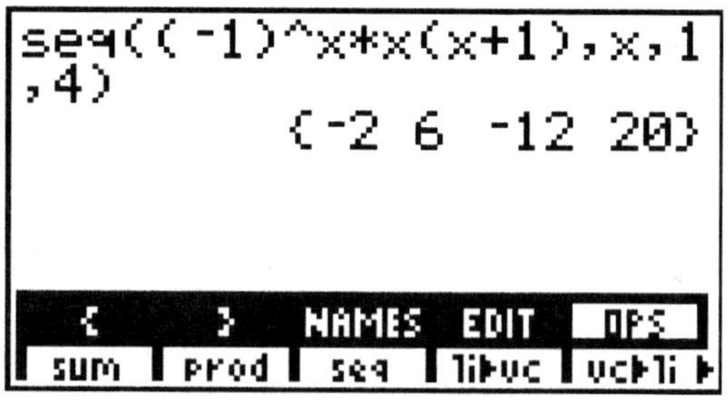

- One way to access the seq(function is to press [2nd] [CUSTOM] to access the CATLG-VARS menu. Press [F1] to select CATALOG.
- Press the [6] key to jump to the S's. Cursor down until you find seq(. Press [ENTER] to put seq(on the Home screen.
- Another way to access the seq(function is to press [2nd] [−] and go to LIST OPS. Press [MORE] [F3] to select seq(. This puts seq(on the Home screen.
- Enter the expression for each sequence, the variable you are using, the beginning and end of the sequence, and finally the increment if needed. Use the [x-VAR] key to enter the variable x.
- To change the sequence to fraction form, press [2nd] [×] to access the MATH menu. Press [F5] MISC [MORE] [F1] to select Frac.
- Repeat the process for the second sequence.
- The screens for each sequence are shown.

Graph the sequence $a_n = \frac{8}{n}$

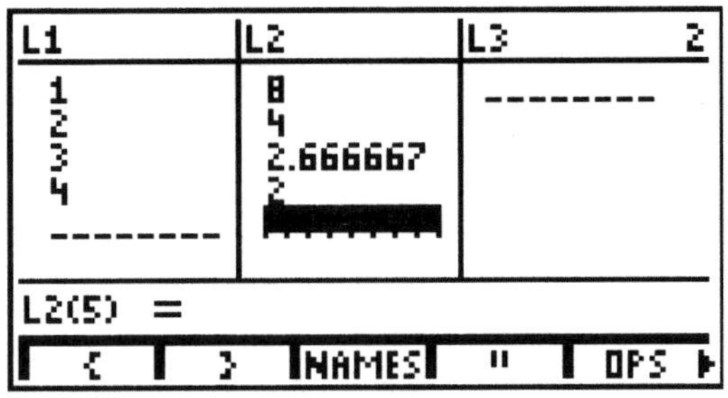

- You must use one of the Plots to graph a sequence on the TI-86.
- Use [2nd] [−] LIST [F4] EDIT to enter the sequence into L_1 and L_2 as shown.
- Press [EXIT] to return to the Home screen.
- Use [2nd] [+] STAT [F3] to see the STAT PLOTS screen.
- Press [F1] to set up PLOT 1 as shown.
- Go to the y= menu. Turn off all other functions and plots.

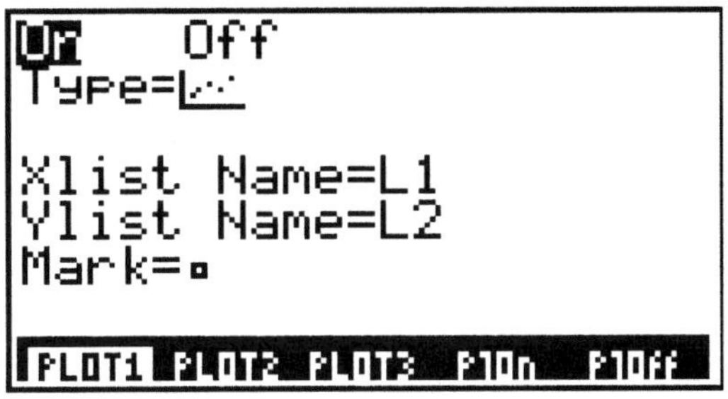

- Set the window for the graph of the sequence as shown. Graph PLOT 1.

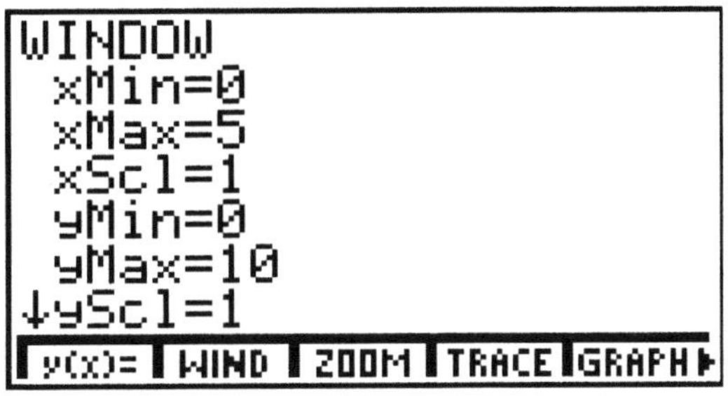

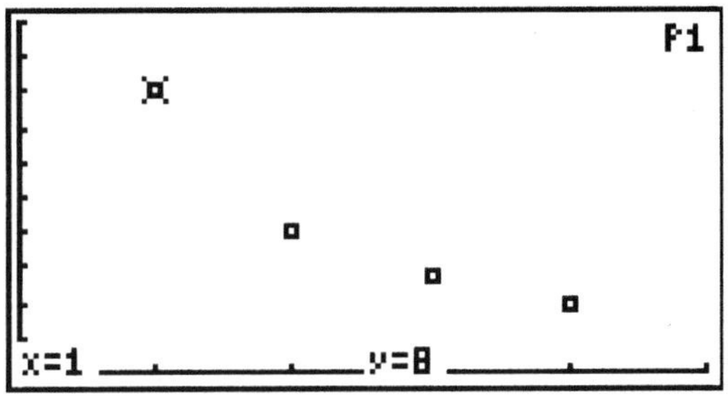

Create a Table of Values for the sequence

The TABLE feature of the TI-86 creates tables from functions only. Since the sequence was entered as a PLOT from two lists, the plot is not treated as a function. Use the lists created above as L_1 and L_2 as the table.

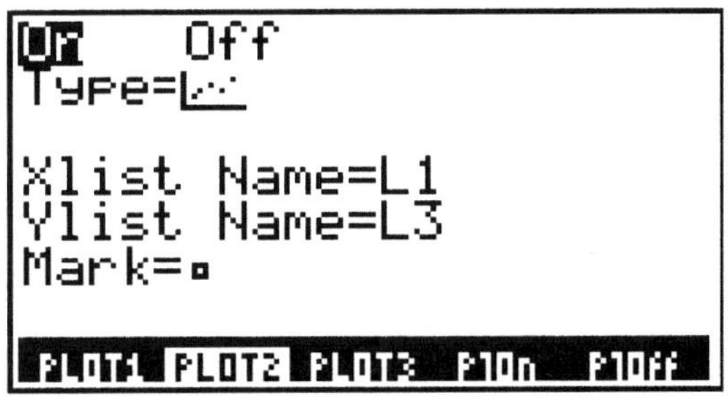

Create the graph and table for $b_n = (-1)^n n(n+1)$

- The only change in the window settings for this sequence are in the *y*-values of the function.
- Set yMin = -20 and yMax = 25.
- Enter the *y*-values of the sequence into L_3.
- Turn off all plots and functions.
- Use Plot 2 to graph the sequence on the TI-86.
- Set up PLOT 2 and graph PLOT 2 as shown.

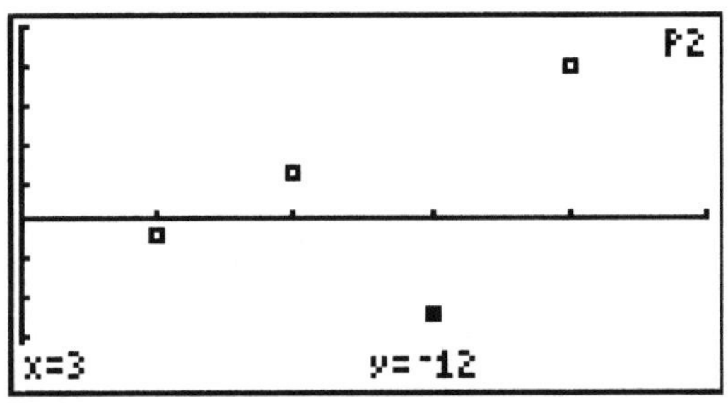